T0254549

Studies in Systems, Decision and Control

Volume 136

Series editor

Janusz Kacprzyk, Polish Academy of Sciences, Systems Research Institute,
Warsaw, Poland
e-mail: kacprzyk@ibspan.waw.pl

The series "Studies in Systems, Decision and Control" (SSDC) covers both new developments and advances, as well as the state of the art, in the various areas of broadly perceived systems, decision making and control– quickly, up to date and with a high quality. The intent is to cover the theory, applications, and perspectives on the state of the art and future developments relevant to systems, decision making, control, complex processes and related areas, as embedded in the fields of engineering, computer science, physics, economics, social and life sciences, as well as the paradigms and methodologies behind them. The series contains monographs, textbooks, lecture notes and edited volumes in systems, decision making and control spanning the areas of Cyber-Physical Systems, Autonomous Systems, Sensor Networks, Control Systems, Energy Systems, Automotive Systems, Biological Systems, Vehicular Networking and Connected Vehicles, Aerospace Systems, Automation, Manufacturing, Smart Grids, Nonlinear Systems, Power Systems, Robotics, Social Systems, Economic Systems and other. Of particular value to both the contributors and the readership are the short publication timeframe and the world-wide distribution and exposure which enable both a wide and rapid dissemination of research output.

More information about this series at http://www.springer.com/series/13304

Umberto Papa

Embedded Platforms for UAS Landing Path and Obstacle Detection

Integration and Development of Unmanned Aircraft Systems

Umberto Papa
Department of Science and Technology
Parthenope University of Naples
Centro Direzionale di Napoli, Is. C4
Naples
Italy

ISSN 2198-4182 ISSN 2198-4190 (electronic)
Studies in Systems, Decision and Control
ISBN 978-3-319-89237-5 ISBN 978-3-319-73174-2 (eBook)
https://doi.org/10.1007/978-3-319-73174-2

Softcover re-print of the Hardcover 1st edition 2018

Printed on acid-free paper

This Springer imprint is published by Springer Nature
The registered company is Springer International Publishing AG
The registered company address is: Gewerbestrasse 11, 6330 Cham, Switzerland

Fortis Fortuna Adiuvat
To my parents, to my wife.
Thank you for your support.

Foreword

The Author Umberto Papa started his career in aeronautics after the graduation in Aerospace Engineering at Seconda Università di Napoli (now Università della Campania "Luigi Vanvitelli"). He had followed my lessons of Flight Mechanics and I had the opportunity to notice his passion for aeronautics.

He was one of the winners of Ph.D. contest in Topographic and Navigation Sciences at Università di Napoli Parthenope and started his Ph.D. studies in 2013; I had kept in my mind his strong enthusiasm and I enjoyed to be designated as tutor. The idea to develop studies about small UAS technologies arose by chance, watching a small quadcopter that was seriously damaged after a hard landing. During the first year, he was involved in the "state of art" about UAS technologies and focused his following activities on sensor data fusion for UAS applications. He was an enthusiastic researcher who spent his time in the Laboratory of Flight and Navigation, marrying theory and practice. His infectious passion was transferred to students and graduates that constituted a small team whose Umberto was leader.

Umberto proposed an application of his Ph.D. thesis for industrial monitoring at a national contest by Finmeccanica (now Leonardo Company), the most important industrial group in Italy; his job was strongly appreciated and he was one of the three winners. Finmeccanica proposed him a six-month stage at Pomigliano d'Arco (Naples-Italy) plant. At the end of this period he was hired by the same Company. Therefore, moving to Leonardo in 2016, he has remained busy in the aeronautical field, covering many aspects of aircraft design and operation including unmanned air vehicle systems.

With such a wide-ranging knowledge of "Mini UASs world", taking into appropriate account his young age, he is skilled to write this book. UAS technologies are quickly evolving, therefore it is very difficult to learn from young engineers and technicians who have been involved in their application; but this book should be very useful reading for everyone working in the field of small UAS systems.

It is strongly recommended to young engineers and students that are planning to research and work in UAS: it is an excellent example of a route that they have to follow if they want achieve good results through modest economic means but driven by a true strong enthusiasm.

Prof. Giuseppe Del Core
Associate Professor, Flight Mechanics
and Aircraft Systems
Parthenope University of Naples
Centro Direzionale di Napoli, Is. C4
Naples, Italy

Preface

The strong passion for aircraft and embedded electronics has been a fundamental driving force for my work and life. The possibility to have the aircraft described in this work, a small autonomous vehicle, at home, and make some experimentations on it, e.g. to analyze different sensors for its safe landing, was a wonderful opportunity: some of the essential findings described in the book were indeed obtained at home!

The advance in electronics and robotics have boosted the application of unmanned (autonomous) vehicles to replace their human (manned) counterpart in the execution of tedious and risky tasks or just as a more cost-effective alternative Unmanned aircraft systems have grown exponentially in the last ten years, appearing in various scenario, from the military to the civil one. There are many classes of unmanned aircraft, and many types within each class, produced by several manufactures. This book covers small UAS only, i.e. with a weight inferior to 10 kg.

It introduces the classes and types of small UAS available, and examines several sensors which can be used for the estimation of the correct landing path and at the same time for detecting any obstacles around the landing zone. The book covers ultrasonic, infrared, and optical (camera module) sensors, discusses their fusion as well as their interaction with the atmosphere. Besides them, other sensors that are described were used on the embedded platform to have information about temperature, humidity, or position (attitude).

Naples, Italy Umberto Papa

Acknowledgements

This book is based on research conducted by the author between 2013 to the present. It has developed out of a series of papers published in journals and lectures given at several conferences. I am grateful to a number of friends and colleagues for encouraging me to start the work, persevere with it, and finally to publish it.

I wish to thank, for their technical and moral support, Prof. Giuseppe Del Core and Prof. Salvatore Ponte, my two supervisors; Prof. Luigi Iuspa, my mentor, and Alberto Greco. I also wish to thank my friends: Salvatore Russo, Giuseppe Martusciello, Paola Di Donato, Halyna Klimenko, Gennaro Ariante, Francesco De Luca, Sonia Basso, Barbara Cruciani, Anna Innac and all the people that have surrounded me in this fantastic life experience. I apologize if I did not name them all, but this would require countless pages.

Finally, I would like to acknowledge the support and infinite love of my family—my parents, Angelina and Salvatore; my sister, Francesca; my brothers-in-law, Claudio, Domenico, Pasquale; my sister-in-law, Loredana; and my little treasure Valeria. Last but not the least my life, Giovanna.

They all kept me going, and this book would not have been possible without them.

Contents

1 Introduction to Unmanned Aircraft Systems (UAS) 1
1.1 Introduction 1
1.2 Why UAS 3
1.3 UAS Classification 3

2 Sonar Sensor Model for Safe Landing and Obstacle Detection 13
2.1 Introduction 13
2.2 System Design 14
2.3 Landing Path Extraction 17
2.4 System Structure 18
2.5 Test Case and Results 21

3 Atmosphere Effects on the SRS 29
3.1 Introduction 29
3.2 Theoretical Framework 29
3.3 SRS Testing 31
3.4 Experimental Analysis Setup and Results 32
3.5 SRS Sensibility 38
3.6 DHT11—Temperature and Humidity Sensor 38

4 Integration Among Ultrasonic and Infrared Sensors 47
4.1 Introduction 47
4.2 Infrared Sensor—IRS 47
4.3 Sensors Integration 52
4.4 Simulation and Results 56
4.5 Conclusion 60

5 Optical Sensor for UAS Aided Landing 63
5.1 Introduction 63
5.2 Vision-Based Embedded System 64
5.2.1 Camera 64

5.3 Extract the Camera Position and Attitude . . . 68
5.3.1 Camera Calibration . . . 69
5.3.2 Point's Extraction . . . 69
5.3.3 Orientation . . . 71
5.4 Validation Procedure . . . 72
5.4.1 Design of Landing Pattern . . . 73
5.4.2 Photogrammetric Survey . . . 73
5.4.3 Camera Orientation . . . 75
5.4.4 Comparison . . . 76
5.5 Results . . . 76
5.6 Conclusion . . . 78

6 UAS Endurance Enhancement . . . 81
6.1 Introduction . . . 81
6.2 HUAS Conceptual Design . . . 82
6.3 Weights Estimation and Balloon Sizing . . . 84
6.3.1 Takeoff Weight Estimation . . . 85
6.3.2 Ballon Static Performance and Sizing . . . 86
6.4 Preliminary Results . . . 89
6.5 Conclusion and Further Work . . . 92

7 Conclusions . . . 97

Appendix . . . 99

References . . . 103

About the Author . . . 107

Author Award . . . 109

Author Publications . . . 111

Chapter 1
Introduction to Unmanned Aircraft Systems (UAS)

1.1 Introduction

Unmanned Aerial Vehicles (UAVs), also denominated Unmanned Aircraft Systems (UAS) by FAA (Federal Aviation Administration), have gained great attention for many applications in the scientific, civil, and military sectors. An UAS can be defined just simply as a system (Valavanis 2008). The system comprises a number of subsystems which include the aircraft, its payloads, the control station(s) (and, often, other remote stations—Ground Station GS), aircraft launch and recovery subsystems where applicable, support subsystems, communication subsystems, transport subsystems, etc.

It must also be considered as part of a local or global air transport/aviation environment with its rules, regulations, and disciplines.

Due to their ability to perform dangerous, sensitive, environmentally critical or dull tasks in a cost-effective manner and with increased maneuverability and survivability, UASs play an important role in various military and civil applications. The numerous military applications include reconnaissance, surveillance, battle damage assessment, and communication relays.

Possible civil applications for UAS include monitoring and surveillance of areas (urban traffic, coast guard patrol, border patrol, detection of illegal imports, archaeological site prospection, etc.), climate research (weather forecast, river flow, etc.), agricultural studies, air composition and pollution studies, inspection of electrical power lines, monitoring gas or oil pipe lines, entertainment and TV, etc. Most civilian uses of UASs require the air vehicle to fly at speeds lower than 50 kts (70 km/h) and at low heights, and many applications need the ability of the aircraft to hover (for example, for power line inspection, subsurface geology, mineral resource analysis, or incident control by police and fire services) (Valavanis 2008).

Historically, the first UAS was introduced during World War I (1914–1919), registering the long participation of the US military with this type of crafts (OSD 2002). These early forms of UASs were discarded by political leaders and military

U. Papa, *Embedded Platforms for UAS Landing Path and Obstacle Detection*,
Studies in Systems, Decision and Control 136,
https://doi.org/10.1007/978-3-319-73174-2_1

experts mainly for their inability to change the battlefield. Only few people, even then, predicted their future potentials.

UASs were also used in Vietnam, in the Desert Storm Operation (1991) and in the conflict of the Balkan Peninsula in the early 1990s. During this period, consensus toward UASs increased.

As early as in 1997, the total income of the UAS global market reached $2.27 billion dollars, a 9.5% increase over 1996.

Figure 1.1 shows the total year (2000) funding of the US DOD (OSD 2001), for UAS and VTOL (Vertical Takeoff and Landing) vehicle design.

The events on 9/11, the war in Afghanistan and in Iraq have changed completely the perception about UASs putting them also on the everyday life map. Nowadays UASs are first subject in media coverage and TV documentaries. As stated in (Dickerson 2007), Europe spends for UASs just about €2 billion, 13% of total US funds.

The VTOL vehicle segment is another evolving sector of the overall UAS market, as shown in Fig. 1.2. Most of VTOLs were used for military application mainly in the US.

The aim of this book is to classify and choose a type of UAS and design an embedded sensors platform useful to assist the remote pilot during the landing procedure.

The design and implementation of an electronic platform were performed by using a bouquet of low-cost sensors (ultrasonic, IR, Optics, etc.) for attitude control and obstacle-sense-and-avoid during the landing procedure at low altitude and low velocity.

Moreover, in some cases (e.g. urban traffic monitoring) this type of platform needs to execute extended missions with significant flight duration time. Increasing endurance generally comes at a cost in terms of fuel consumption and airframe complexity, resulting in reduced efficiency of payload and/or range for size, mass, and financial cost. Chapter 6, at the end of this work, will explain a possible solution in order to increase the UAS endurance.

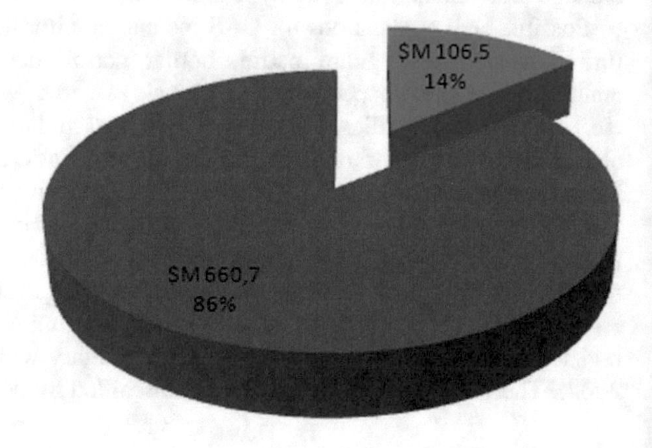

Fig. 1.1 US government funds in UASs/VTOLs year 2000

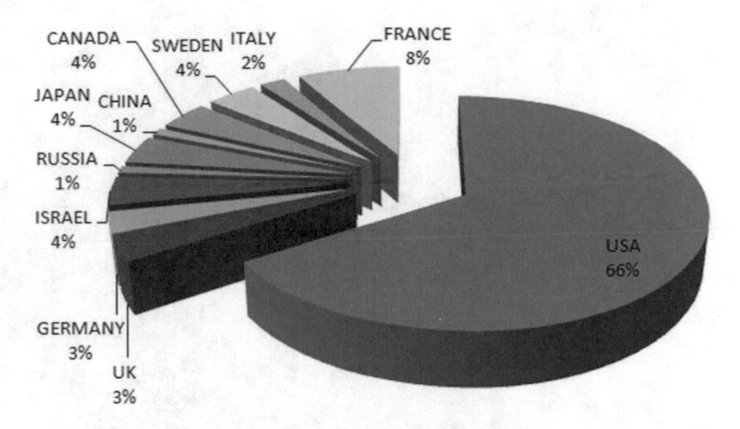

Fig. 1.2 Percentage of VTOL models produced over the world

1.2 Why UAS

As stated in Valavanis (2008), OSD (2002, 2005), Valavanis et al. (2008a, b), UASs are preferred over manned aircraft mainly because the human cost of losing lives if the mission is unsuccessful is avoided, but also because unmanned vehicle has better and sustained alertness over humans during monotonous operations.

This is relevant in civilian applications, for example, urban traffic, coast guard patrol, border patrol, detection of illegal imports, archaeological site prospection, etc.), climate research (weather forecast, river flow, etc.), agricultural studies, air composition and pollution studies, inspection of electrical power lines, monitoring gas or oil pipe lines, entertainment, and TV, etc. In this case, the advantages are justifying their preference over their manned counterparts.

The statement that UASs are best suited for "dull, dirty and dangerous" missions has merit and it is supported because:

- Dull—Long operations, which require more than 30- or 40- h missions, are ideal for UASs involvement. This type of missions (in particular their duration), if manned, can compromise proficiency and functionality of crewmembers.
- Dirty—In case of chemically contaminated areas, the UASs alternative prevails. This solution will be fundamental for flying into nuclear clouds (as appended in Černobyl' disaster in April 26, 1986) avoiding humans usage.
- Dangerous—Operations like reconnaissance over enemy territory may result in loss of human lives, thus UASs are preferred.

1.3 UAS Classification

In modern times, UASs appeared during World War I (1914–1919). However, the idea of a flying machine was conceived about 2500 years ago.

Fig. 1.3 Leonardo da Vinci aerial screw, the first UAS in the history

Autonomous mechanism and first aircrafts (flying machines) were studied and designed by Leonardo da Vinci (1452–1519), the first engineer. Da Vinci, in 1483 designed an aircraft that was able to hover, like today's helicopter does. This first aircraft was called aerial screw or air gyroscope (Fig. 1.3).

This machine is considered the ancestor of today's helicopters.

Currently, a broad range of UASs exists, from small and lightweight fixed-wing aircrafts to rotor helicopters, large-wingspan airplanes and quadrotors, each one for a specific task, generally providing persistence beyond the capabilities of manned vehicles (Valavanis 2008).

According to Table 1.1 (Eisenbeiss 2004), UASs can be categorized with respect to mass, range, flight altitude, and endurance.

This classification, based on radius of action in operation, has been replaced by newest terms (Austin 2010), as follows:

- Long-range UAV—replaced by HALE and MALE;
- Medium-range UAV—replaced by TUAV;
- Close-range UAV—often referred to as MUAV or midi-UAV.

These new acronyms are explained in Table 1.2:

Some UAS, and related properties are depicted (Figs. 1.4 and 1.5).

Table 1.1 Extract of UAV categories defined by UVS (Unmanned Vehicle Systems Association) international

Category name	Mass (kg)	Range (km)	Flight altitude (m)	Endurance (h)
Micro	<5	<10	250	1
Mini	<25/30/150	<10	150/250/300	<2
Close range	25–250	10–30	3000	2–4
Medium range	50–250	30–70	3000	3–6
High altitude, long range	>250	>70	>3000	>6

Figures in the category "mini UAV" depend on different countries

Table 1.2 Categories of systems based upon air vehicle types (Austin 2010)

Category	Range	Endurance (h)
HALE—*High Altitude Long Endurance*	Altitude over 15,000 m	24+
MALE—*Medium Endurance Long Endurance*	Altitude 5000–15,000 m	24
TUAV—*Medium Range or Tactical UAV*	Horizontal distance 100–300 km	
Close-range UAV	Horizontal distance about 100 km	
MUAV or *Mini UAV*	About 30 km and a weight below 20 kg	

Fig. 1.4 Northrop Grumann—RQ-4 Global Hawk HALE

Length	13.5 m
Wingspan	35.4 m
Height	4.6 m
Empty weight	3850 kg
Max takeoff weight	10400 kg
Powerplant	1 × Turbofan Allison Rolls Royce AE3007H
Cruise speed	575 km/h
Range	22779 km
Endurance	32+ h
Service ceiling	20,000 m

Fig. 1.5 General Atomics—RQ-1 Predator MALE

Length	8.22 m
Wingspan	14.8 m
Height	2.1 m
Wing area	11.5 m^2
Empty weight	512 kg
Max takeoff weight	
Powerplant	1 × Rotax 914F turbocharged four-cylinder engine
Cruise speed	130–165 km/h
Range	1100 km
Endurance	24 h
Service ceiling	7,500 m

The US Department of Defence (DoD) also has classified UASs into five categories as shown in Table 1.3 (US Army 2010).

Considering Table 1.2, some Large UASs in service were depicted in Figs. 1.6, 1.7, 1.8 and 1.9.

VTOL (Vertical Takeoff and Landing) aircrafts provide many advantages over Conventional Takeoff and Landing (CTOL) vehicles. Most notable are the capability of hovering in place and the small area required for takeoff and landing.

Among VTOL aircrafts such as conventional helicopters and crafts with rotors like the tilt-rotor and fixed-wing aircrafts with directed jet thrust capability, the quadcopter, or quadrotor (a helicopter with four rotors fixed on the ends of a cross-shaped frame), is very frequently chosen, especially in the academic research on mini or micro-size UAVs, as an effective alternative to the high cost and complexity of the conventional rotorcrafts, due to its ability to hover and move without the complex system of linkages and blade elements present in a standard single-rotor vehicle (Nonami et al. 2010, this chapter). Employing four rotors to create differential thrust, the quadrotor gains flexibility, swift maneuverability, and increased payload. A useful comparison of different types of VTOL miniature flying robots (MFR) can be found in Nonami et al. (2010). Table 1.4, adapted from

Table 1.3 UASs classification according to the US Department of Defense (DoD)

Category	Size	Maximum takeoff weight (kg)	Normal operating altitude (m)	Airspeed (km/h)
Group 1	Small	0–9.9	<365 AGL[a]	<185.2
Group 2	Medium	10–24.9	<1066.8	<463
Group 3	Large	<599	<5486.4 MSL[b]	<463
Group 4	Larger	>599	<5486.4 MSL	Any airspeed
Group 5	Largest	>599	>5486.4	Any airspeed

[a]*AGL* Above ground level
[b]*MSL* Mean sea level

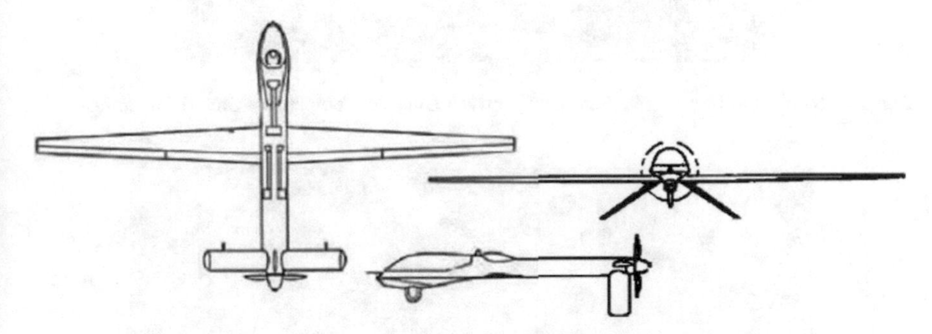

Fig. 1.6 The MQ-1 Predator built by General Atomics Aeronautical Systems Inc

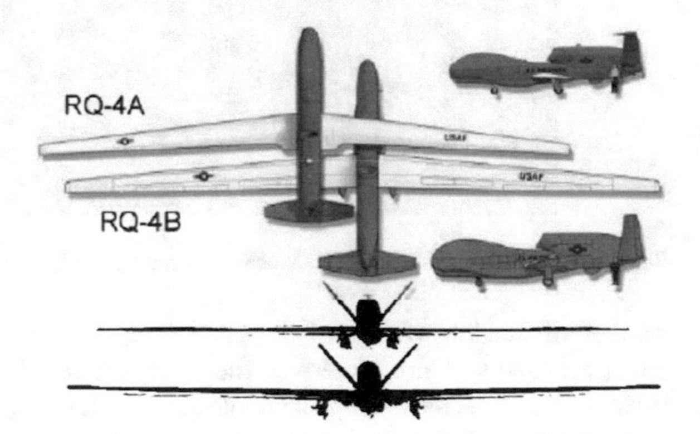

Fig. 1.7 The RQ-4 A/B Global Hawk designed by Northrop Grumman

(Austin 2010), gives quality indexes (from 1 = bad, to 4 = very good) for some design issues pertaining to different VTOL vehicle concepts, namely, bird-like (A), single-rotor (B), tandem rotors (C), insect-like (D), axial rotor (E), blimp (F), coaxial rotors (G), quadrotor (H).

The quadrotor has a good ranking among VTOL vehicles, yet it has some drawbacks. For example, the craft size is comparatively larger, energy consumption is greater, therefore providing short flight time, and the control algorithms are very complicated, due to the fact that only four actuators are used to control the six

Fig. 1.8 Sky-Y built by Alenia Aermacchi (now Leonardo Company—Aircraft Division)

Fig. 1.9 The RQ-7A/B Shadow 200 manufactured by AAII

degrees of freedom (DOF) of the craft (a quadrotor is a typical example of an under-actuated system). Moreover, the changing aerodynamic interference patterns between the rotors have to be taken into account (Austin 2010). Unlike planes, there is no rudder, no ailerons, just propellers, typically brushless. The only way to modulate flight is by spinning the rotors at different speeds.

Light UASs quadrotors use plastic propellers, which resist breaking on impact because they are flexible and safer. Some models use carbon fiber or other more rigid materials (planes frequently use wood or nylon/glass). Carbon fiber propellers are dangerous, even deadly, and should be used only by experienced pilots and well away from people. Unless extreme performance is a concern, the benefits of carbon fiber over plastic are marginal on multirotors. For this type of UASs, the clockwise/counterclockwise pairing of every two motors is important.

Table 1.4 Comparison among different VTOL concepts and design issues (adapted from (Valavanis 2008), see text for coefficient explanation)

Design driver	A	B	C	D	E	F	G	H
Mechanics simplicity	1	1	1	1	3	4	3	4
Aerodynamics complexity	1	1	1	1	1	3	1	4
Low-speed flight	2	4	3	2	3	4	4	4
Stationary flight	1	4	4	2	4	3	4	4
Control cost	2	1	2	1	1	3	4	3
Payload/volume	2	2	3	1	2	1	4	3
Maneuverability	3	4	3	3	2	1	2	3
High-speed flight	3	2	2	3	4	1	1	3
Miniaturization	2	2	2	4	3	1	4	3
Power cost	3	2	2	3	2	4	2	1
Survivability/endurance	2	1	1	3	3	3	3	1
Total quality index	22	24	24	24	28	28	32	33

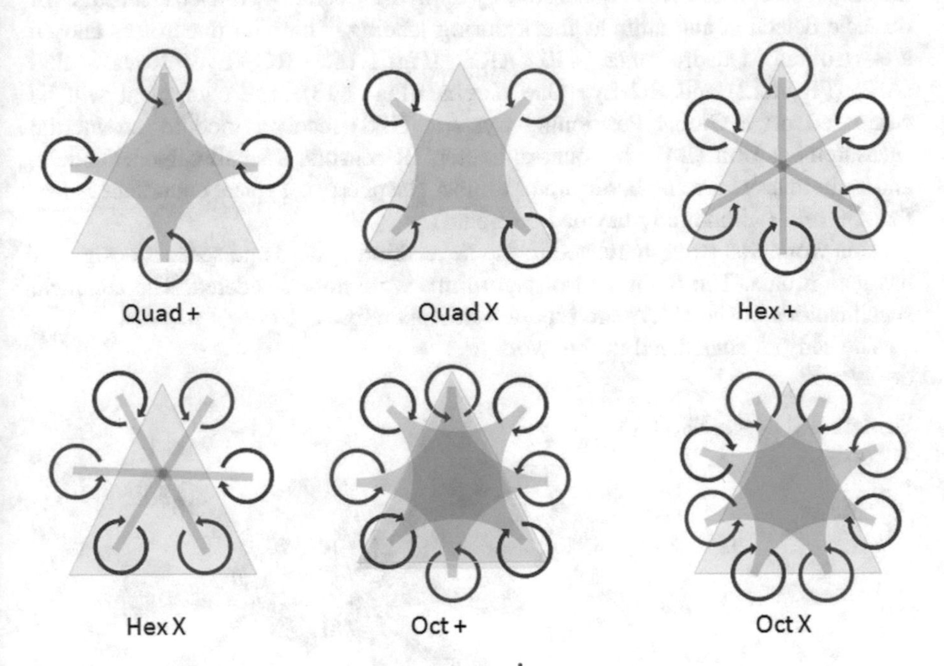

Fig. 1.10 Possible quadrotor configurations

In Fig. 1.10 are depicted some arrangements for a number of quadrotor types.

The UAS heading is represented by the top vertex of blue triangle (Fig. 1.11), and the circular symbols represent the clockwise (red) and counter clockwise (blue) of the propellers' rotation.

Fig. 1.11 Conrad 450 (ArF) 35 MHz

This work proposes to design and improve a safe landing system for a non-expensive, commercial quadrotor, by applying a series of low-cost sensors, for obstacle detection and attitude check during landing. The mini quadrotors chosen are: Conrad Quadrocopter 450 ARF (Fig. 1.11), RC Eye NovaX 350 (ArF) (Fig. 1.12) and RC Eye One Xtreme (Fig. 1.13). The equipment will be composed of: a Global Positioning System (GPS) receiver module, an Inertial Measurement Unit (IMU), a sonar altimeter, IR sensors, a small camera module, and main microcontroller hardware (Arduino, Raspberry). Typical endurance is less than 30 min (without any payload on board).

This work was strongly related to the acquisition system and sensors design for navigation data. The flight control algorithms were not considered. The technical specifications of the UAVs are reported in Appendix A.

The sensors considered in this work are:

Fig. 1.12 RC Logger NovaX 350 ArF

Fig. 1.13 RC Eye One Xtreme

- Ultrasonic Sensor (SRS)—Sonar Sensor Model for Safe Landing (Chap. 2) and Atmosphere Effects Correction (Chap. 3);
- Infrared Sensor (IRS)—Integration among Ultrasonic and Infrared Sensor (Chap. 4);
- Optical Sensor (OPS)—Safe Landing procedure through Raspberry Pi Camera using photogrammetry algorithms (Chap. 5).

Chapter 6 describes a new concept of UAS, in order to increase its endurance during the flight. In Chap. 7 draws conclusions and addresses further work on the investigated topics.

Chapter 2
Sonar Sensor Model for Safe Landing and Obstacle Detection

2.1 Introduction

In UASs (Unmanned Aerial Systems) applications, it is important to find and track all scenario properties (e.g., obstacles). Performing obstacles and terrain avoidance from an UAS platform is challenging for several reasons (Valavanis 2007). The UAS limited payload and power available give significant limitations on the total size, weight, and power requirements of potential sensors. Embedded sensors systems like LIDAR and RADAR are typically too large and heavy for the UAS.

The sonar sensor tracking is one of the best ways to detect any obstacle in a flight zone like bats or dolphins do. In short, a signal time delay, measured from the ultrasonic echo pin, gives distance measurements considering the speed of sound. The fundamental approach is to calculate the distance from the delay of the ultrasonic burst.

In civilian applications, UASs are mostly used due to their low cost and size. The cost of these UASs partially depends on the embedded sensors used for the flight control (IMU, GNSS, etc.).

However, in some conditions, such as urban or low altitude operations, the GNSS receiver antenna is prone to losing the line-of-sight from the satellite, therefore, making the GNSS receiver unable to deliver high-quality position information. This is quite dangerous for closed-loop control systems during the landing. Therefore, in these cases, an ultrasonic sensor (SRS—Sonar Ranging Sensor) may be very useful for altitude control. Moreover, SRSs require lower computational efforts to provide target distance than camera-based systems and have higher energy efficiency than laser ones.

Furthermore, obstacle detection is an important task for fully autonomous UASs. Ultrasonic proximity sensors are a good compromise in terms of cost, energy efficiency,and accuracy for low-distance obstacle detection and altitude control.

The goal of this chapter is to design and evaluate an SRSs system useful for assisting UASs during landing. The system should be able to detect any obstacles

U. Papa, *Embedded Platforms for UAS Landing Path and Obstacle Detection*,
Studies in Systems, Decision and Control 136,
https://doi.org/10.1007/978-3-319-73174-2_2

on the landing field. At the same time, the system maps the ground, and if the tilt angle of the field is greater than 30°, the landing procedure is aborted. The range of altitude considered is 20–150 cm from the ground (landing field).

2.2 System Design

The overall system is composed of four ultrasonic sensors HC-SR04 (HC-SR04 2010), one microcontroller Arduino Mega 2560 (Arduino.cc 2016), cables and a standing structure. The system concept is depicted in the Fig. 2.1.

The ultrasonic sensor module HC-SR04 (Fig. 2.2) provides a distance datum range of 2–400 cm no-contact measurement function, the range of accuracy can reach up to 3 mm. The module includes ultrasonic transmitters, receivers, and a control circuit. The shield has four pins (V_{cc}, Ground, Echo, and Trigger) (Table 2.1).

In order to make measurements with both ultrasonic sensors, the following procedures (HC-SR04 2010) have been carried out (see Fig. 2.3):

- Send a pulse signal of at least 10 μs (HLT—High-Level Signal) on the trigger pin for HC-SR04;
- The ultrasonic transceiver automatically sends eight 40 kHz ultrasonic waves to the target and detect whether there is a pulse signal back;
- A pulse waveform is transmitted on Echo and IO pins. The pulse width is comparable with the time of fly.

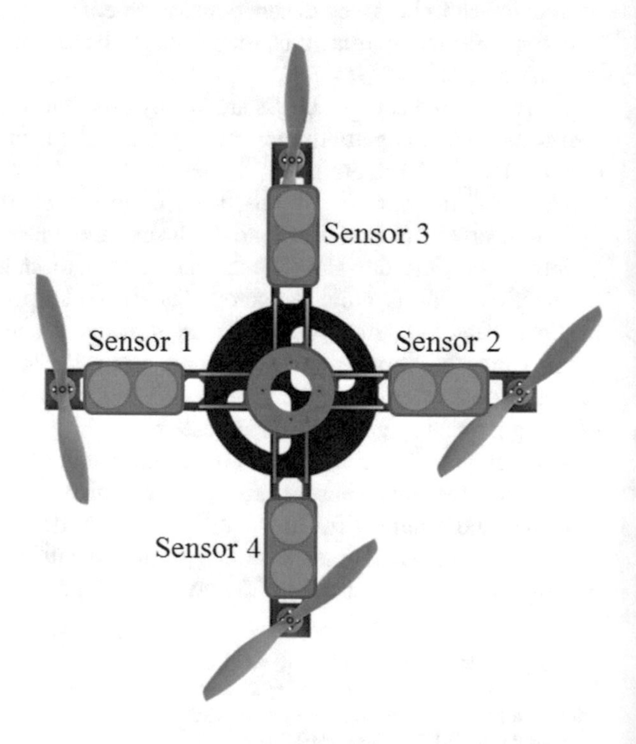

Fig. 2.1 Model structure concept

Fig. 2.2 HC-SR04 Ultrasonic Sensor Module front and back view details

Table 2.1 HC-SR04 Technical Chart (HC-SR04 2010; Ping Parallax 2016)

	HC-SR04	Parallax Ping)))
Supply voltage (V DC)	5	5
Supply current	15 mA	30 mA; 35 mA max
Range (cm)	2–400	2–300
Input trigger	10 μs TTL pulse	2 μs min, 5 μs typ.
Echo pulse	Pos. TTL pulse	115 μs–18.5 ms
Burst frequency	40 kHz	40 kHz for 200 μs
Measuring angle	<17° for side	<20° for side
Dimension (mm)	45 × 20 × 15	45.7 × 21.3 × 16

The distance was given by:

$$d = \frac{(HLT * a)}{2} \tag{2.1}$$

where *HLT* is high-level time and *a* is the speed of sound (340 m/s in the air at 20 ° C). If no obstacle is detected, the output pin will give a 38 ms high-level signal.

Ultrasonic sensors work at 5 V DC and they were linked to a microcontroller Arduino (Arduino.cc 2010; Timmins 2011) through the four digital pins for HC-SR04.

The microcontroller platform was the single board Arduino Mega 2560 (Fig. 2.4).

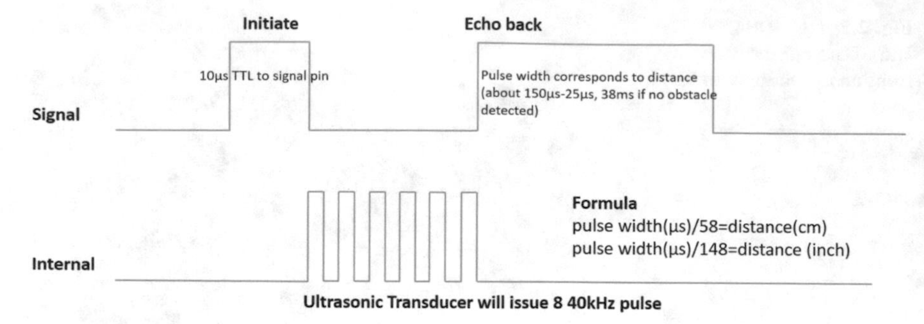

Fig. 2.3 Sequence chart

Fig. 2.4 Arduino Mega 2560 microcontroller (Arduino.cc 2016)

The Arduino Mega 2560 Microcontroller is based on the ATmega2560. It has 54 digital input/output pins (of which 15 can be used as PWM outputs), 16 analog inputs, 4 UARTs (hardware serial ports), a 16-MHz crystal oscillator, a USB connection, a power jack, an ICSP header, and a reset button. It contains components needed to support the microcontroller; simply connect it to a computer with a USB cable or power it with an AC/DC adapter or battery to get started.

Arduino acquires data distances from ultrasonic sensors and sends them to a PC laptop, via USB interface. This operation can be performed also via Wireless (Bluetooth or Wi-Fi), thanks to a specific module (Fig. 2.5) added to Arduino board.

The wiring between the SRS and Arduino was very simple, as depicted by the electrical scheme in the figure below.

Figure 2.6 considers the SRS coupled with an Arduino Uno, but the wiring remains the same also for Arduino Mega.

Fig. 2.5 Arduino Wireless and SD shield, together used respectively for a wireless communication and for storing data

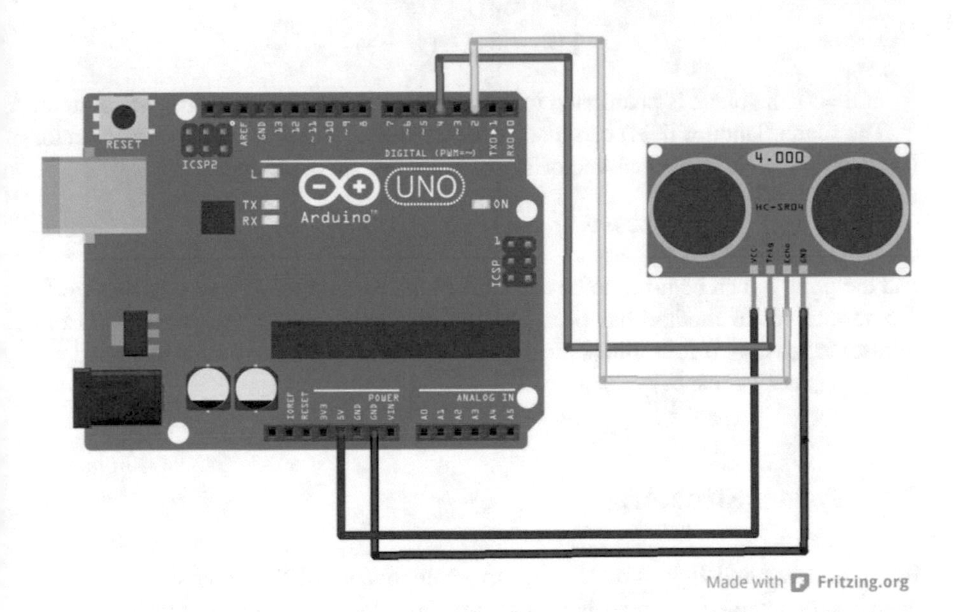

Fig. 2.6 Electrical wiring scheme between SRSs and Arduino Uno. The scheme was made with Fritzing Software

2.3 Landing Path Extraction

The four sensors can extract distance measurements, useful for landing fields glide estimation.

In general, a plane is described by an equation of the form $ax + by + cz + d = 0$, just solving Eq. (2.2) with the Laplace rule:

$$\begin{bmatrix} (x-x_1) & (y-y_1) & (z-z_1) \\ (x_2-x_1) & (y_2-y_1) & (z_2-z_1) \\ (x_3-x_1) & (y_3-y_1) & (z_3-z_1) \end{bmatrix} = 0 \tag{2.2}$$

so:

$$a(x-x_1)+b(y-y_1)+c(z-z_1)=0 \tag{2.3}$$

where the coefficient a, b, and c are:

$$\begin{aligned} a &= \begin{bmatrix} (y_2-y_1) & (z_2-z_1) \\ (y_3-y_1) & (z_3-z_1) \end{bmatrix} \\ b &= \begin{bmatrix} (x_2-x_1) & (z_2-z_1) \\ (x_3-x_1) & (z_3-z_1) \end{bmatrix} \\ c &= \begin{bmatrix} (x_2-x_1) & (y_2-y_1) \\ (x_3-x_1) & (y_3-y_1) \end{bmatrix} \end{aligned} \tag{2.4}$$

if $d = 0$, the plane is positioned in the origin of the Cartesian coordinate system.

The plane (landing field) can also be described by the point and normal vector (Eq. 2.5). A suitable normal vector is given by the cross product:

$$n = (p_2 - p_1) \times (p_3 - p_1) \tag{2.5}$$

and the point p_0 can be taken to be any of the given points p_1, p_2, or p_3. In this work, the normal vector method has been utilized, where p_1, p_2, and p_3 are points from ultrasonic sensors. If four ultrasonic sensors are considered, the points extracted are p_1, p_2, p_3, and p_4 respectively.

2.4 System Structure

For a compact and light structure, all the components of the acquisition systems have been contained in an ad hoc platform. In order to extract the four (or three) distance points, a cross configuration of the sensors has been chosen (Fig. 2.7). The distance between sensors is the minimum, in order to avoid any interference between them.

The four ultrasonic sensors, positioned in a cross configuration, the microcontroller and the battery have been installed in the structure as shown in Fig. 2.8.

The rendering in Fig. 2.8 considers the material property and the real dimensions (scale 1:1) of the objects, in order to evaluate if any interactions between them exist, before the assembly.

Fig. 2.7 The CAD model of the SRSs holder, made in ABS with a 3D printer, where each position contains one SRS

Fig. 2.8 Whole SRSs system rendering built, thanks to a CAD software. In this figure, the previous ABS SRSs holder and the electronic devices (Arduino, LCD display, and battery) were depicted

The final structure has been realized thanks to a 3D printer, and for simplicity of manufacturing, the whole structure was divided into various subcomponents. Figure 2.9 shows the printing of the arm, which contains the ultrasonic sensor.

The material used for these cover structures was plastic ABS (Acrylonitrile Butadiene Styrene). They were printed by a 3D printer (3DRAG 2016) available in the Laboratory of Navigation. This hardware can print objects of maximum size of

Fig. 2.9 SRS holder during print process with the 3DRAG printer in the Air Navigation Laboratory of the University of Naples "Parthenope", Napoli, Italy

$20 \times 20 \times 20$ cm using ABS or PLA (Polylactic acid or polylactide) 3 mm wires. The printer uses the X/Y for printing plane and Z for the cart; this particular configuration allows simplifying the extrusion system, which no longer has to move on a horizontal axis and is simply fixed to the structure that moves along the Z-axis (3DRAG 2016).

The complete system is depicted in the figure below (Fig. 2.10).

To get a feedback in real time of the standalone system, an LCD (16×2) display was mounted. Furthermore, a micro SD card shield (Fig. 2.11) was installed

Fig. 2.10 SRSs system developed, including the wiring. In this case, the wireless shield and the SD shield were in two different shields and not in the configuration of Fig. 2.5

Fig. 2.11 SD card shield used for SRSs data storing

in order to have the system data stored; in case of wireless shield (Fig. 2.5) fault. The overall system weight was about 250 gms.

The system mounted on the UAV is depicted in Fig. 2.12:

2.5 Test Case and Results

For the experimental data collection, a range of 20–160 cm was considered; a typical range covered during a small UAS landing procedure (Farid bin Misnan et al. 2012). The system (cyan box in Fig. 2.13) was initially tested in static sessions and positioned on a tripod.

The distance extracted depends on the speed of sound, defined as follows:

$$a = \sqrt{\gamma RT} \tag{2.6}$$

where γ is the adiabatic index, R is the molar gas constant (approximately 8.3145 J/mol•K), and T is the temperature (K). The speed of sound depends on the transmission material (e.g., air, water, wall, etc.) and temperature. This aspect is dealt with in Chap. 3. In this case, the test was conducted in ISA (International Standard Atmosphere) conditions, see Table 2.2.

The data collected was useful to determinate the HC-SR04 performance, in terms of mean, standard deviation, and variance. Preliminary results of data distances, acquired from COM8 port (Fig. 2.14) and computed in Matlab®, are shown below:

The data results confirm an accuracy of the ultrasonic sensor of about 3 mm, as specified (HC-SR04 2010) (Fig. 2.15).

Successively, it has been possible to extract the LPTA (Landing Plane Tilt Angle) from the four ultrasonic sensors (HC-SR04) acquisition; considering Eq. (2.5) (Fig. 2.16).

The system autonomously decides to land or not, evaluating the LPTA and if there were any obstacles around.

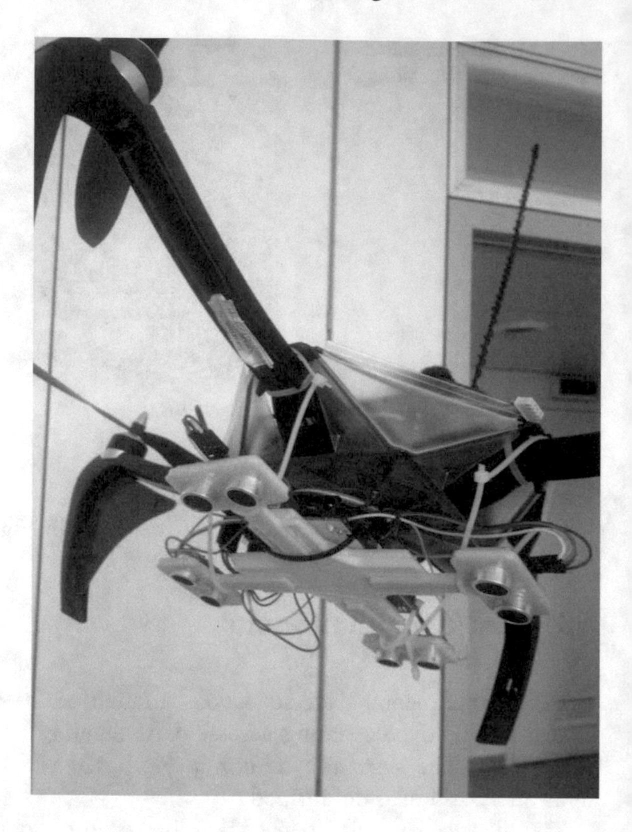

Fig. 2.12 SRSs system mounted below Conrad 350 ArF, used during preliminary tests

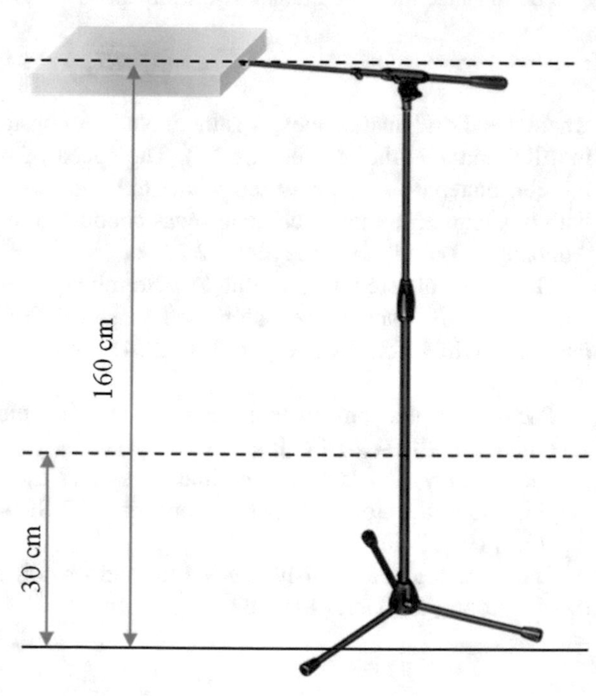

Fig. 2.13 System setup for the static sessions in the Flight Dynamic Laboratory

Table 2.2 International Standard Atmosphere model, based on average conditions at mid-latitudes, as determined by the ISO's TC 20/SC 6 technical committee. It has been revised from time to time since the middle of the twentieth century (Graham 2006)

International Standard Atmosphere ICAO	
Temperature MSL[a]	15 °C—288.15 K
Pressure MSL[a]	101,325 Pa
Density	1.225 kg/m^3
Humidity	0%

[a]*MSL* Mean Sea Level

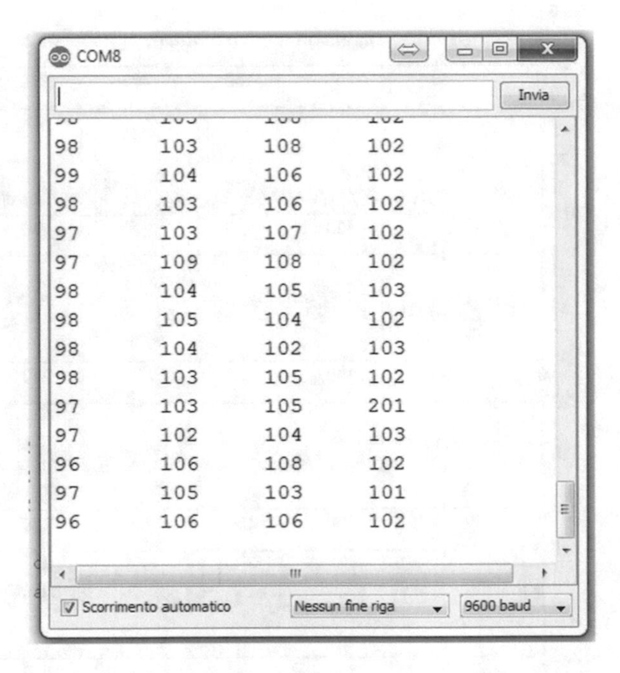

Fig. 2.14 Four HC-SR04 distances from serial port (COM8) displayed in the Arduino Serial Monitor

In order to cover a wide area, another configuration of ultrasonic sensors was considered. This configuration works like a RADAR, the ultrasonic sensor rotates, thanks to a servomotor, in order to inspect a wide area.

In addition, the holding structure was made in ABS (Fig. 2.17), designed with a 3D CAD software and printed through 3D printer.

In Fig. 2.18, the scanned area contains a box (20 × 20 × 20 cm) inside, so in this case, the landing was not permitted, because the box's height is greater than UAS's legs. When an obstacle is detected, as in this case, the user is alerted through an acoustic beep or a blinking led (Fig. 2.19).

By coupling the ultrasonic sensor with a servomotor, it was possible to generate a map (mapping) in real time like a radar. Considering this hypothesis, a radar display was made, utilizing Processing (a software similar to Arduino IDE),

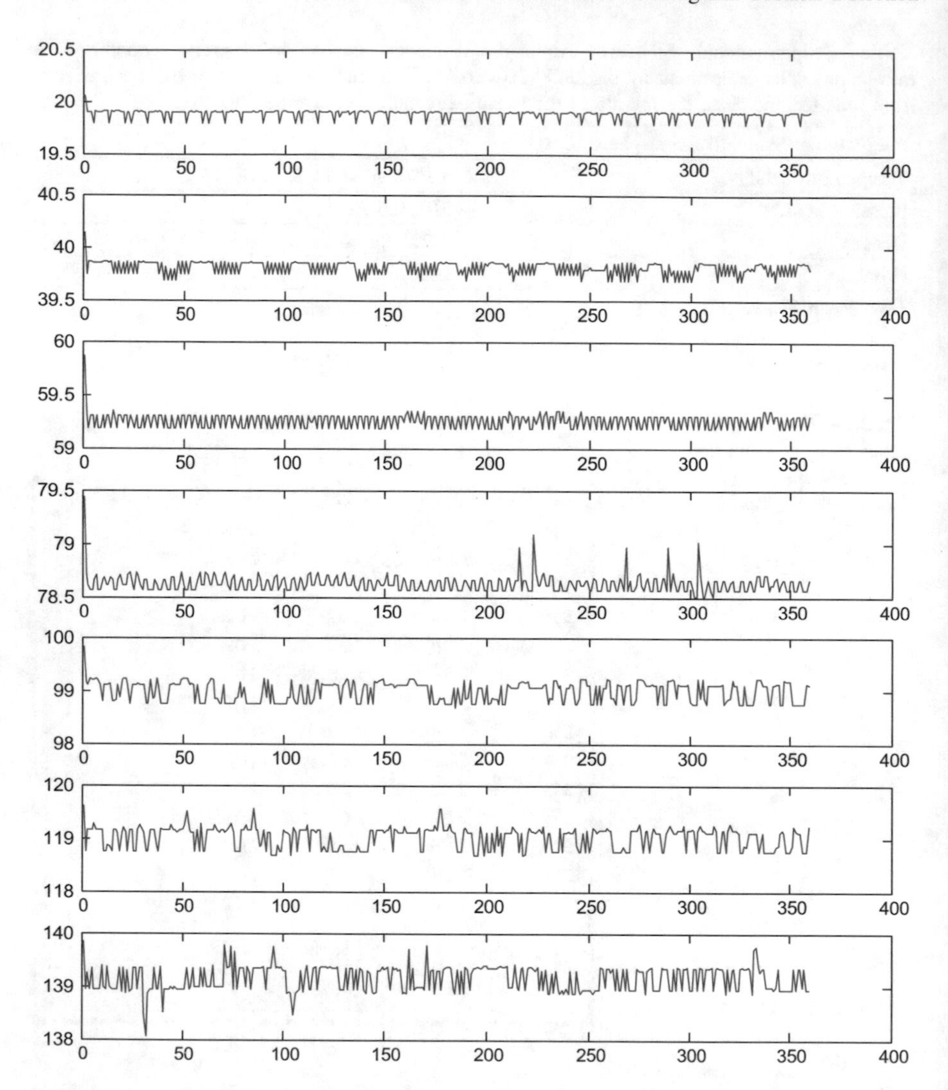

Fig. 2.15 Data collection of distances from 20 to 150 cm, with SRS HC-SR04

generating for the user on the GS (ground station) an interactive interface that can detect and show any obstacles in the scanned area, in real time. Figure 2.20 shows ultrasonic radar window.

The data collected from the sensors have been processed by the microcontroller, and by means of the Matlab® environment, the performance (in Matlab®) and abilities on the arrangements of the HC-SR04 used for UAS application mapping were analyzed. The data from the experimental session was directly transmitted, via Bluetooth, to a GS where a computational process was needed in order to

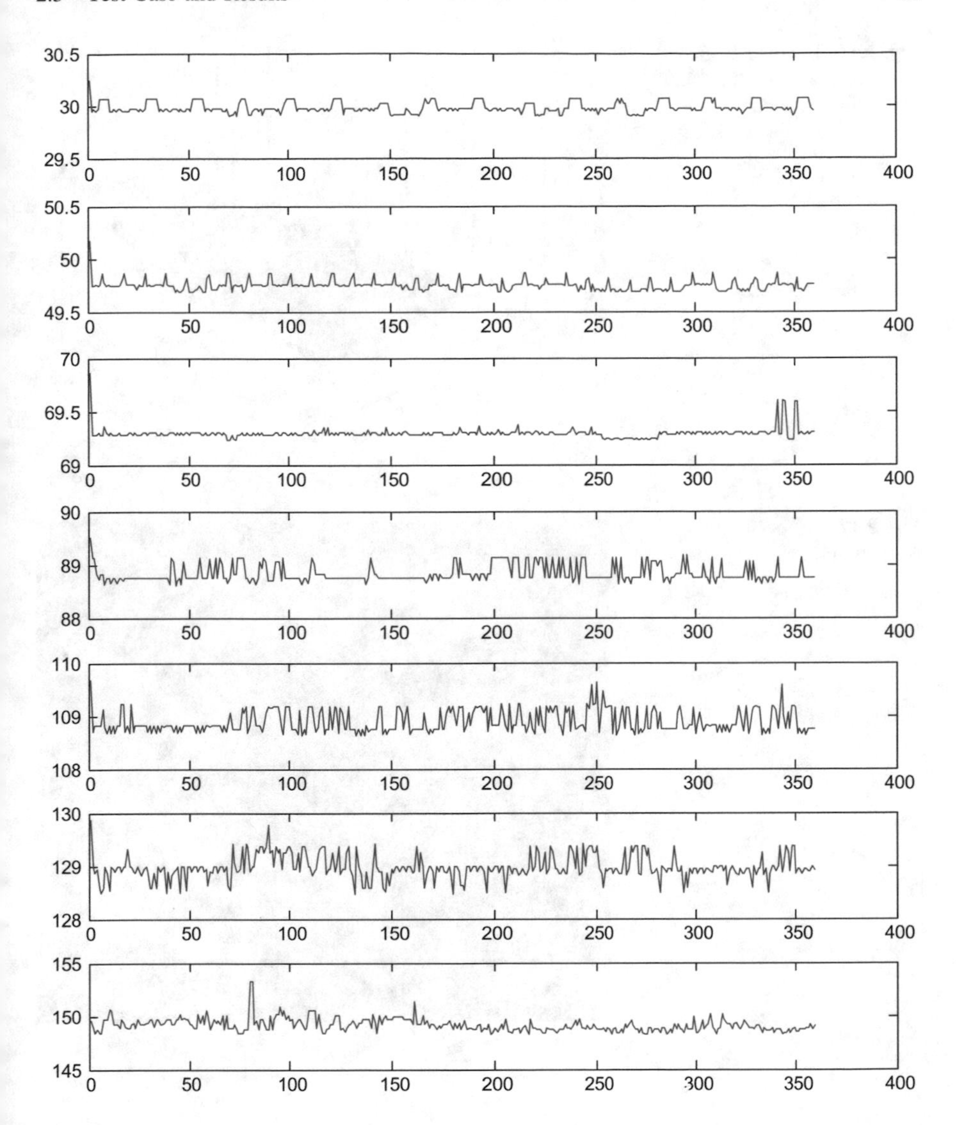

Fig. 2.15 (continued)

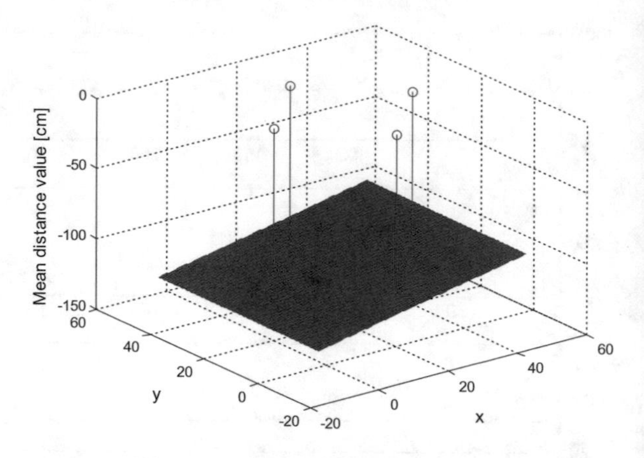

Fig. 2.16 Landing Plane extrapolation from the four SR04 distances data

Fig. 2.17 HC-SR04 rotary configuration

Fig. 2.18 Obstacle (box) detected in the scanned area

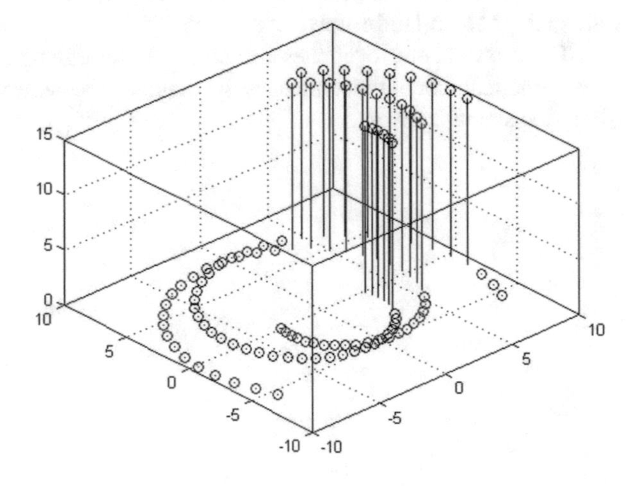

Fig. 2.19 Landing field mapping, obstacle detected

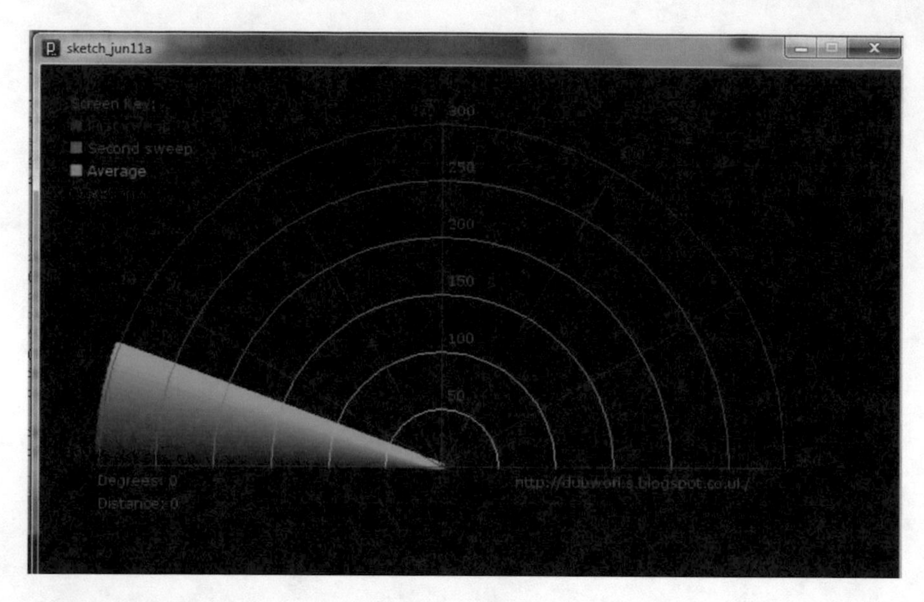

Fig. 2.20 Ultrasonic Radar window developed in Processing IDE

implement the mapping procedure. This work was useful in order to estimate and validate the SR sensor (HC-SR04) performance, and build a preliminary system that aids the UAS to land safely.

The next chapter evaluates the atmospheric effects on HC-SR04, generating a mathematical model that enhances the distance measurements extraction, in order to avoid systematic errors.

Chapter 3
Atmosphere Effects on the SRS

3.1 Introduction

Ultrasonic sensors are a good compromise in terms of cost, energy efficiency, and accuracy for low distance obstacle detection. Various fields of interest could find this book interesting, for example, structural damage inspections in critical areas (e.g., L'Aquila earthquake, Italy, or hurricane Ike in Galveston, Texas, US).

Unfortunately, the measurements provided by ultrasonic sensors are affected by systematic errors due to relative humidity and atmospheric conditions. Previous works have dealt with studies concerning specular reflections on ultrasonic sensors (Yi et al. 2000), but they did not consider temperature and relative humidity interactions.

The aim of this chapter is to analyze the effects of temperature and relative humidity conditions on distance measurements provided by ultrasonic sensors, during landing or hovering. The investigations on two commercial ultrasonic sensors (HC-SR04 by Electrofreaks and Parallax PING) took place in a small environmental chamber, model KK-105 CH. In order to avoid the systematic errors, two mathematical models have been carried out for both relative humidity and temperature effect compensations.

3.2 Theoretical Framework

The distance measurements provided by SRS are based on the measurements of the time of fly of ultrasonic waves reflected on a target. As mentioned in the previous chapter, and confirmed by the Eq. (2.6), the speed of sound depends on the temperature and humidity.

Knowing the numerical value of the sound speed, it is possible to evaluate the distance between the sensor and the target with the following equation (Dean 1979):

U. Papa, *Embedded Platforms for UAS Landing Path and Obstacle Detection*,
Studies in Systems, Decision and Control 136,
https://doi.org/10.1007/978-3-319-73174-2_3

$$d = \frac{c \cdot t_{fly}}{2} \tag{3.1}$$

where t_{fly} is the time interval between the transmission of the ultrasonic wave and the received echo wave and c is the speed of sound.

Equation 3.1 shows that the distance depends linearly on the speed of sound. If the speed of sound grows more than 10% starting from a nominal value, in the distance measurement occurs an error of 10%. This error can be fatal when finding obstacles during UAS flight.

For a real gas, the speed of sound depends on temperature, pressure, and molecular composition. In classical mechanics, the speed of sound is given by:

$$c = \sqrt{\frac{K}{\rho}} \tag{3.2}$$

where K is the modulus of the bulk elasticity of a gas obtained with the multiplication of the adiabatic index γ and the pressure p, while ρ is the density. If we use the ideal gas law to replace p with nRT/V, and replacing ρ with nM/V, the speed of sound for an ideal gas (c_{ideal}), considering the Eq. (3.2), is given by:

$$c_{ideal} = \left[\frac{(\gamma RT)}{M}\right]^{1/2} \tag{3.3}$$

where R is the molar gas constant (approximately 8.3145 J/mol K), T (K) is the absolute temperature, γ is the adiabatic index (1.4 for dry air 273.15–473.15 K) and M_{air} is the molar mass of the gas (for dry air about 0.029 kg/mol).

For the mixture air, the molar mass is the average of the mole fractions of its components and their molar masses. If relative humidity changes, there will be a variation of water particles concentration in the air mixture, so that the molar mass changes. This formulation is valid only for small perturbations on the climate condition.

Based on previous equations (Dean 1979), the speed of sound can be derived according to the following formula:

$$c_{air} = 331.3 + 0.606 \cdot \vartheta \tag{3.4}$$

where ϑ is the temperature in Celsius and 331.3 is the speed of sound (m/s) in dry air (0% humidity) at 0 °C. This formula is derived from the first two terms of the Taylor Expansion:

$$c_{air} = 331.3 \cdot \left(1 + \frac{\vartheta}{273.15}\right)^{1/2} \tag{3.5}$$

An equivalent form Eq. (3.5) is obtained by multiplying the right-hand side by $(273.15)^{1/2}$:

$$c_{air} = 20.05 \cdot (\vartheta + 273.15)^{1/2} \tag{3.6}$$

The equations mentioned, show how the distance measurements depend on temperature and gas molar mass (Dean 1979; Endoh et al. 2003). Hence, it is important to compensate the effect of temperature, relative humidity, pressure, and airflow masses during the distance measurements.

3.3 SRS Testing

The sonic ranging sensors (SRS), or ultrasonic sensors, involved in this analysis are two low-cost commercial models: (a) HC-SR04 and (b) Parallax PING))).

The first one is described in detail in Chap. 2. The second one, Parallax PING))) (Ping Parallax 2016) (Fig. 3.1b) has a similar range of distance measurement, about 2–300 cm.

The shield is quite similar to HC-SR04 Main difference between the SRSs is related to the pins configuration, the PING Parallax consists of three pins (V_{cc}, Ground, and IO), where IO includes trigger and echo tasks on one channel. Table 3.1 shows ultrasonic sensors features, in order to evaluate the differences between them.

The procedure for distance extraction was explained in the previous chapter.

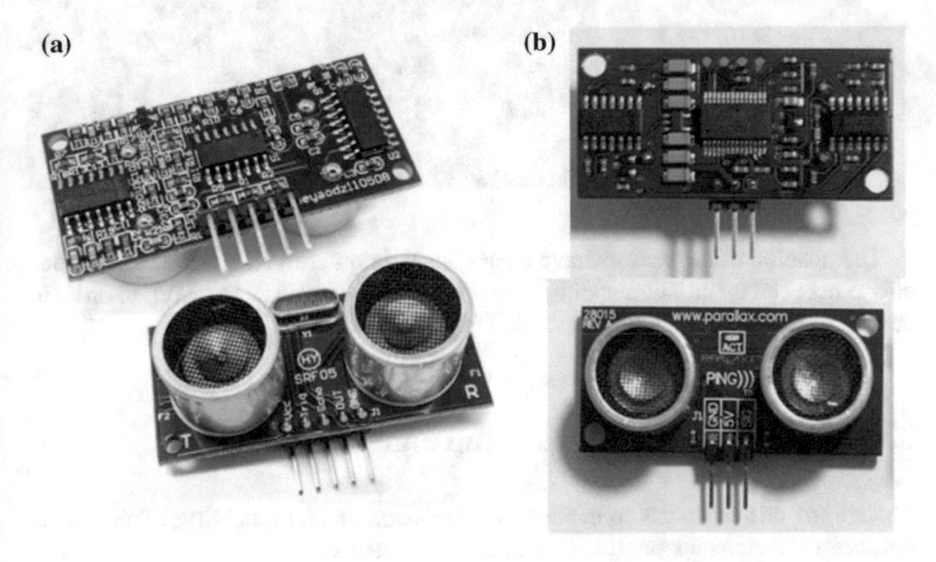

Fig. 3.1 **a** HC-SR04, **b** Parallax PING)))

Table 3.1 SRSs technical chart, according to (Ping Parallax 2016; HC-SR04 2010)

	HC-SR04	Parallax PING)))
Supply voltage	5 V DC	5 V DC
Supply current	15 mA	30 mA; 35 mA max
Range	2–400 cm	2–300 cm
Input trigger	10 μs TTL pulse	2 μs min, 5 μs type.
Echo pulse	Pos. TTL pulse	115 μs to 18.5 ms
Burst frequency	40 kHz	40 kHz for 200 μs
Measuring angle	<17° for side	<20° for side
Dimension	45 × 20 × 15 mm	45.7 × 21.3 × 16 mm

Fig. 3.2 Concept of the structure

The position of the sensors have been chosen in order to reduce the interference effects between them. The standing bar and the base were made in wood, in order to have a structure much light and stable (Figs. 3.2 and 3.3).

3.4 Experimental Analysis Setup and Results

The aim of this section is to evaluate the temperature and humidity influences on distance measurements provided by ultrasonic sensors.

Fig. 3.3 Detail: HC-SR04 and Arduino installed on the wood arm

To investigate these influences, a measurement procedure has been performed in the small size environmental chamber Kambic KK-105 CH (Fig. 3.4) with a temperature range from +5 to 180 °C and humidity range from 10 to 98% (Climatic Chamber KK-105 2016).

Fig. 3.4 KK-105 CH climate chamber used in this work

Technical data of the chamber are reported in the following (Table 3.2).

The reference measurement system for temperature and humidity measurements was a certified probe (HD2817T 2011) model HD2817T (Fig. 3.6). In particular, the Leica DISTO™ D3 (Fig. 3.5) has been used as a reference system for distance measurements.

The tests were made at the LESIM (Laboratorio di Elaborazione Segnali ed Informazioni di Misura—Measurement Information and Signal Processing Laboratory) laboratory at the Department of Engineering of the University of Benevento "Unisannio" (Benevento, Italy).

First, different cases were considered, in order to analyze the ultrasonic sensor's performance. Table 3.3 shows the analyzed range, in terms of temperature and relative humidity, for a given set of distances, namely 19, 23, 28, 33, and 38 cm, obtaining for each subset a (4 × 4) matrix.

All the data were managed through a.vi file (National Instruments 2003) that let the user configure the procedure parameters. Moreover, the application allows

Fig. 3.5 Leica DISTO™ D3

Table 3.2 KK-105 CH technical data (Climatic Chamber KK-105 2016)

External dimensions (WxHxD) (mm)	725 × 1500 × 845	Uniformity @50 °C	±0.5 °C (RH = 50%)
Internal dimensions (WxHxD) (mm)	490 × 500 × 430	Control	MPC-PID
Volume (l)	105	Power supply	230 V 50/60 Hz (±10%)
Humidity range	10…98%	Wattage (W)	3000
Humidity stability	±3%	Interface	RS 232
T&Rh display resolution	0.1 °C/1%	Weight (kg)	~205
Set point resolution T/Rh	0.1 °C/1%	Shelve	1 included
Temperature range (°C)	5…+180	Access port	1 × ϕ 50 mm included
Heating rate	~2.2 °C/min		
Cooling rate	~ 1.5 °C/min		
Temperature range with humidity control (°C)	+10 °C…+95		

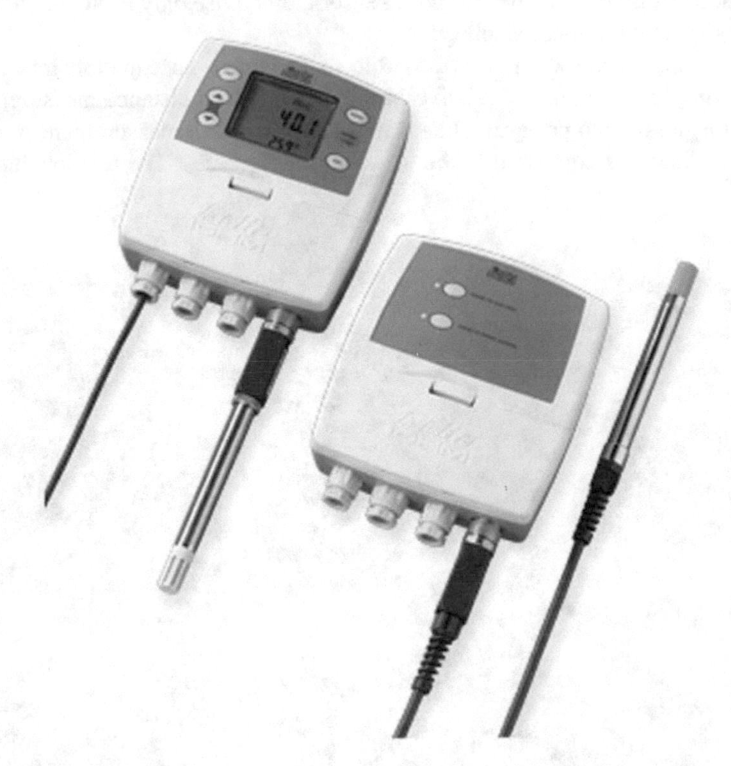

Fig. 3.6 HD2817T, transmitter, indicator, ON/OFF regulator, temperature, and humidity datalogger

Table 3.3 Data value setup

Temperature (°C)	Humidity (%)
10	10
20	30
30	50
40	70

communicating with the microcontroller and the collected data are stored and displayed on the screen thanks to a user-friendly graphic interface (Fig. 3.7).

Afterward, it is also possible to plot, through Matlab® all collected data during the simulation. Considering Fig. 3.8, it is possible to see how the distance measurements provided by Parallax PING))) depend on temperature and relative humidity for a fixed set distance of 28 cm.

In the same way, Fig. 3.9 shows the results of the SR04 sensor.

In this case, a distance of 28 cm has about 2 cm (7%) of variation for a given temperature range (10–40 °C), whereas for relative humidity range (10–70%) the distance variation was smaller.

Variations were quite similar for both sensors, otherwise only PING parallax has shown higher measurement stability.

After all these analyses, it was possible to formulate an equation that allows compensating temperature and relative humidity effects on distance measurements. At first, it is possible to perform a linear regression for each measurement provided by the tested sensors compared to the given distance values. The relationship is as follows:

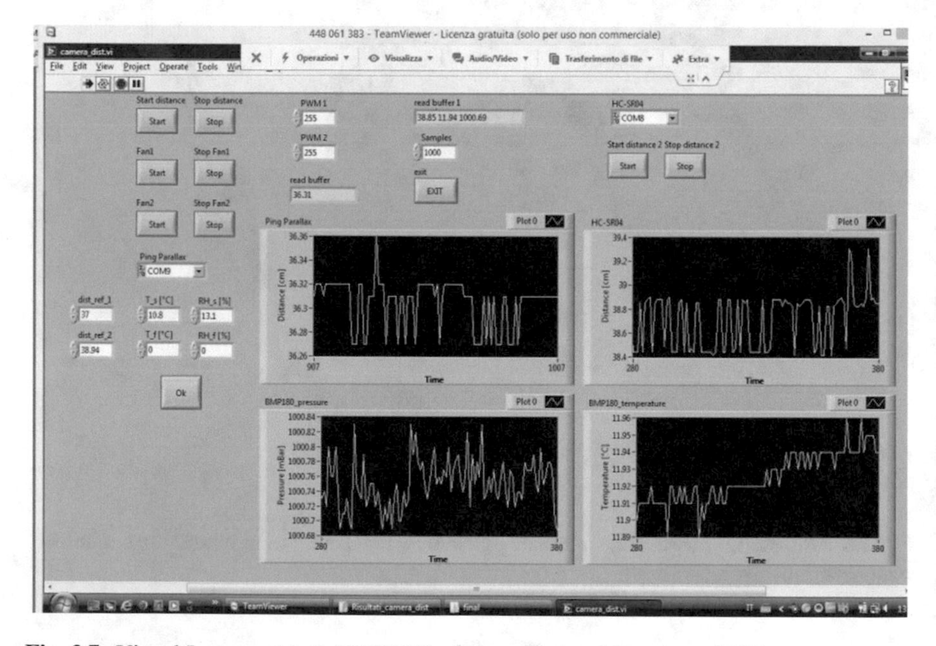

Fig. 3.7 Virtual Instrument in LabVIEW™ window (National Instrument 2003)

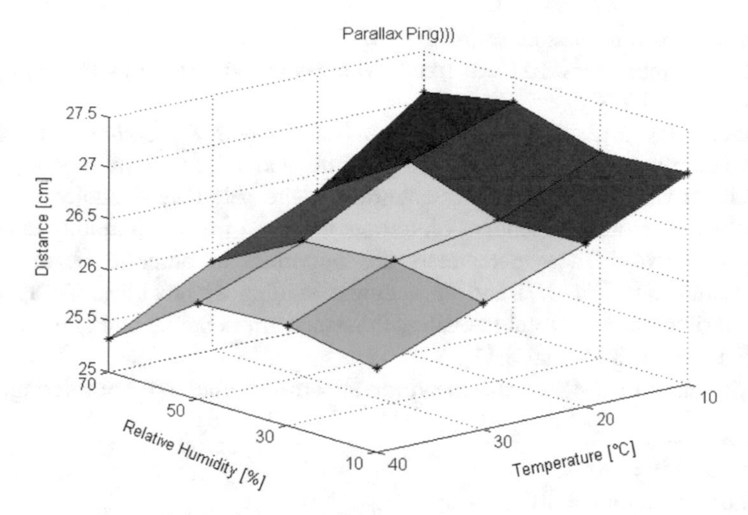

Fig. 3.8 Data changing for PING))) Parallax

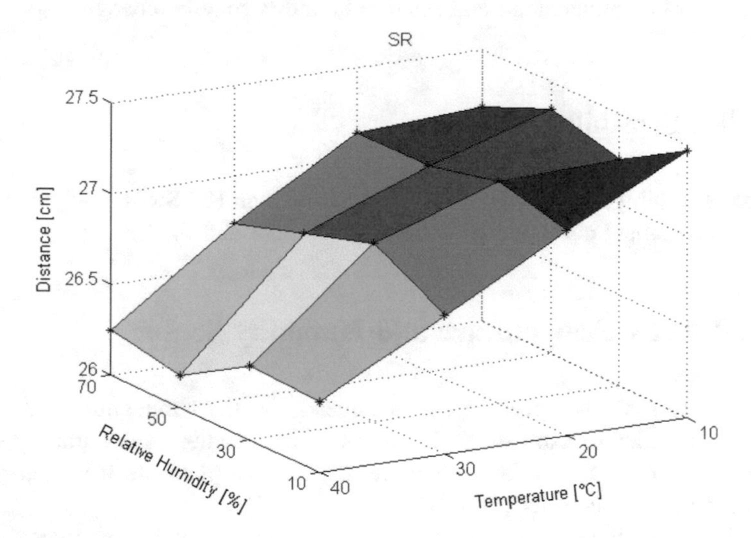

Fig. 3.9 Data changing for HC-SR04

$$y = ax + b \tag{3.7}$$

where a and b were respectively the scale factor and the bias coefficient. These parameters depend on the environmental values of temperature and humidity that affect the speed of sound and its formulation (Eq. 2.6).

The figures below show the results obtained from the two sensors concerning relative humidity (10%) and temperature (equal to 10 °C). The variation of reference distance was compared to the obtained distance, and the trend of the valued scale factor and bias were used to estimate a distance compensation (Eq. 3.7).

Similar procedure was done for the HC-SR04 sensor.

In the previous cases, RH was fixed, whereas in the next cases the temperature was fixed ($T = 10$ °C).

Figures 3.10, 3.12, 3.14, and 3.16 bias coefficients for both sensors do not change linearly with reference to temperature and relative humidity variations. Therefore, it was important to pay attention to the parameters' choice (scale and bias factor), as closely dependent on temperature and relative humidity calculated during the survey. These parameters are important in order to obtain a good compensation law (Eq. 3.7) and an accurate landing or hovering. To appreciate work contribution, it is useful to evaluate distance errors before and after correction (Figs. 3.11, 3.13, 3.15, and 3.17).

Maximum error distance and mean are shown in Table 3.4, considering:

- Distance = 28 cm;
- Temperature = 20 °C;
- Relative humidity = 30%.

Equation 3.7 correction was important to obtain a better distance measurement, in real time when temperature and relative humidity may be changing.

3.5 SRS Sensibility

In terms of stability, the PING))) was much stable than HC-SR04, as figures below show for previously discussed conditions (Figs. 3.18 and 3.19).

3.6 DHT11—Temperature and Humidity Sensor

As previously said, the speed of sound depends on the temperature and relative humidity. In order to estimate their values, a temperature and humidity sensor (Fig. 3.20) (DHT11 sensor, 2010) was installed on the platform. It is managed by the Arduino microcontroller (Fig. 2.4).

This sensor includes a resistive-type humidity measurement component and an NTC (Negative Temperature Coefficient) temperature measurement component, and connects to a high-performance 8-bit microcontroller, offering excellent quality, fast time response, anti-interference ability, and a reduced cost.

The DHT11 communicates through four pin (V_{cc}, Data, NC, and GND) to the Arduino microcontroller (Fig. 3.20).

The sensor waits for a low level and then a high level on its data line. Once it detects a low level that lasts at least 18 ms and then a transition to a high level again, it initializes its transmission circuitry and pulls the data pin low for 80 μs as an indication that it will now start sending data. Then it starts the transmission of the readings. It is a total of 40 bits and each bit (1 or 0) is transmitted as a pulse of

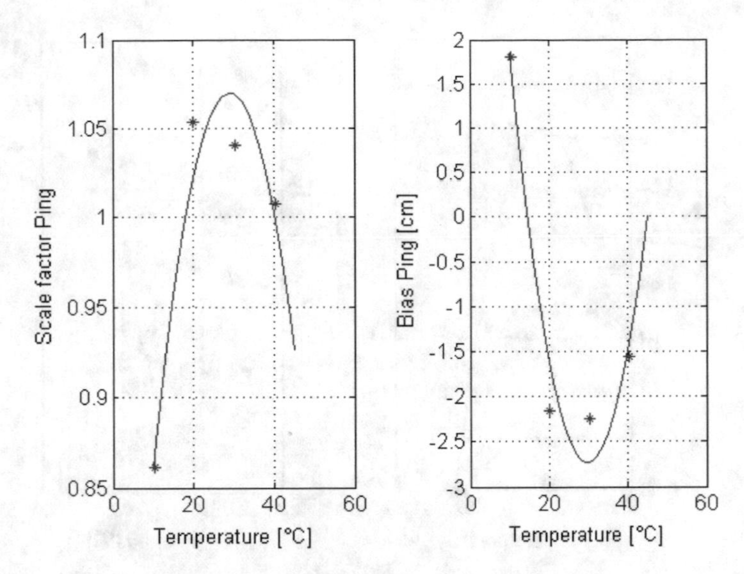

Fig. 3.10 Scale factor and bias (RH = 10%—PING)

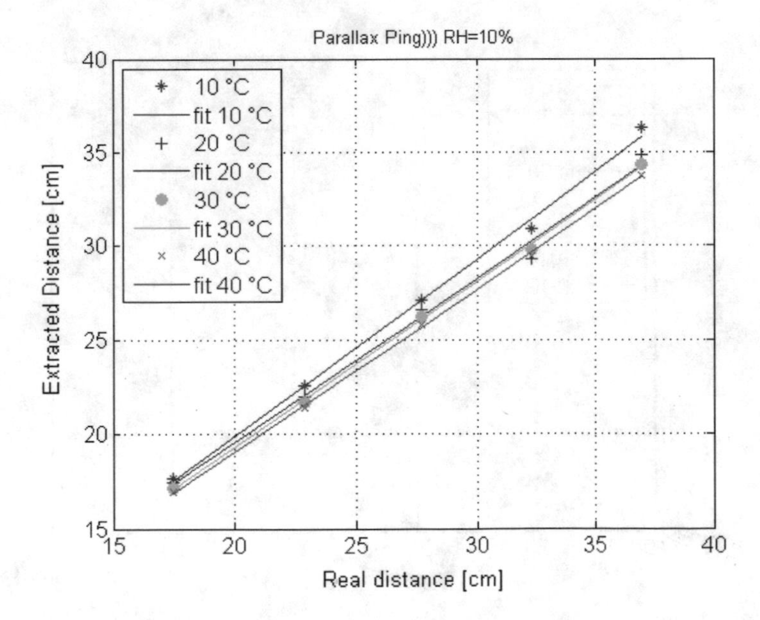

Fig. 3.11 Linear function for correction (RH = 10%—PING)

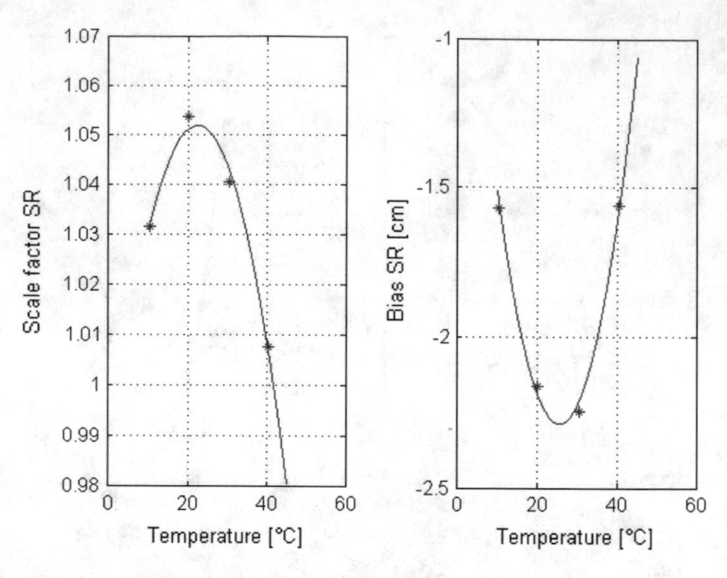

Fig. 3.12 Scale factor and bias (RH = 10%—HC-SR04)

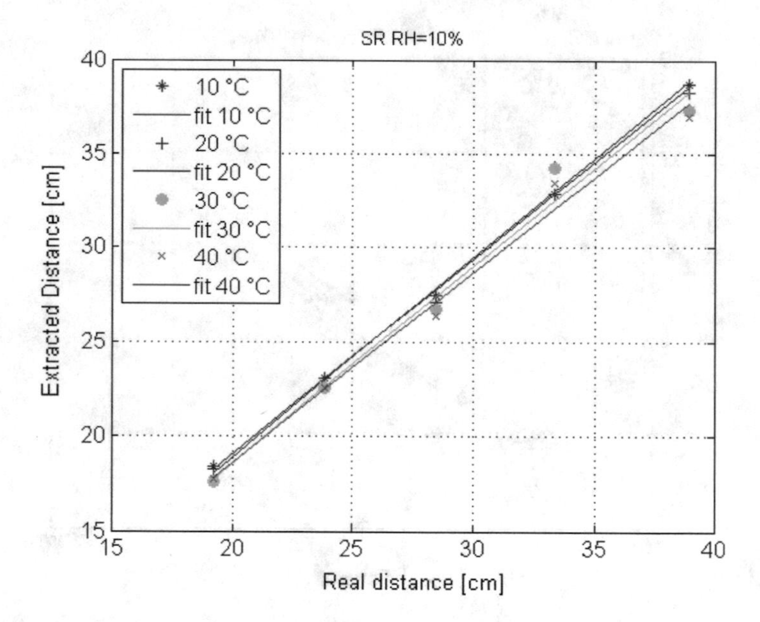

Fig. 3.13 Linear function for correction (RH = 10%—HCSR04)

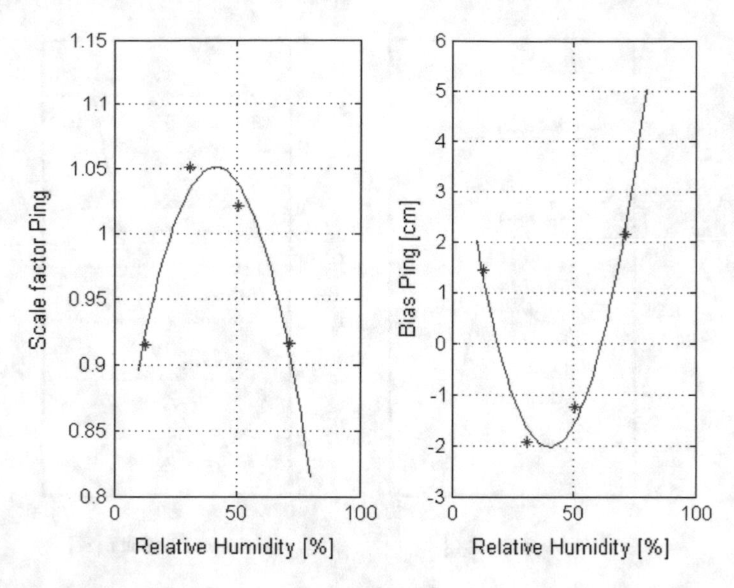

Fig. 3.14 Scale factor and bias (T = 10 °C—PING)

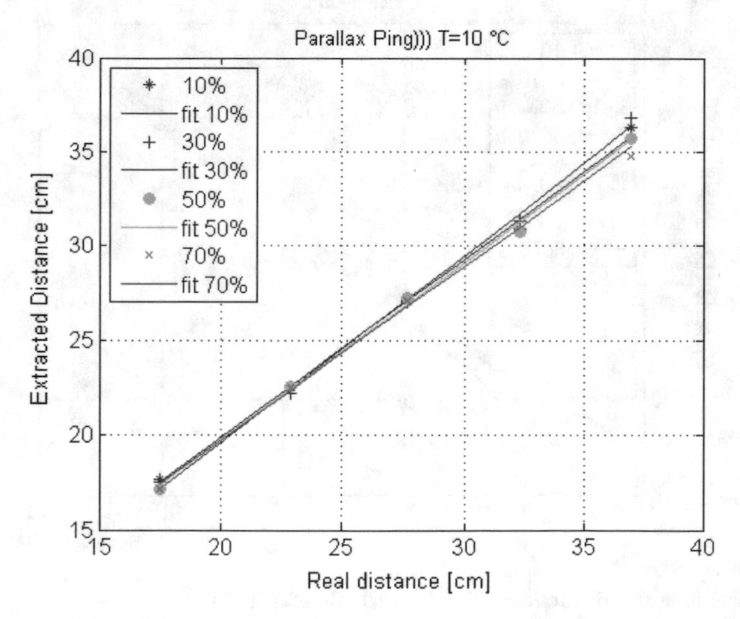

Fig. 3.15 Linear function for correction (T = 10 °C—PING)

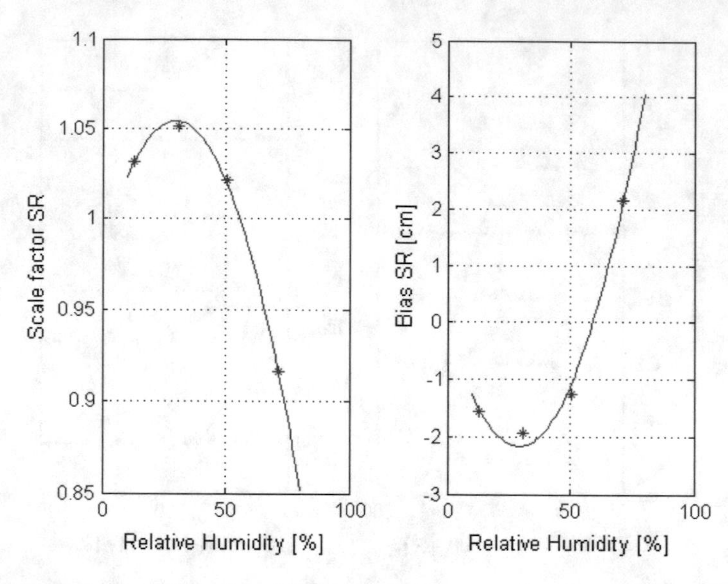

Fig. 3.16 Scale factor and bias (T = 10 °C—HC-SR04)

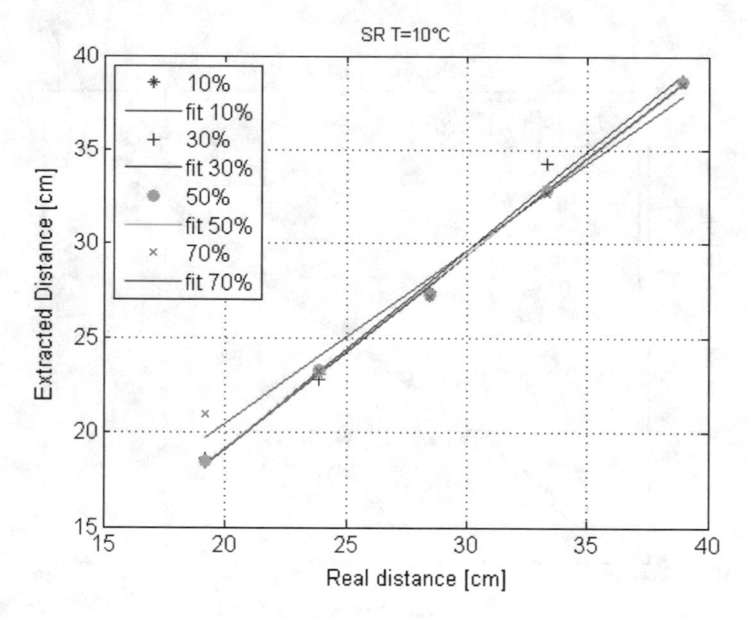

Fig. 3.17 Linear equation for correction (T = 10 °C—HC-SR04)

Table 3.4 SRSs comparison before and after the correction

		maxErr (cm)	Mean (cm)
Ping)))	Not-corrected	1.37	26.57
	Corrected	0.88	27.88
SR04	Not-corrected	0.51	27.22
	Corrected	0.21	28.02

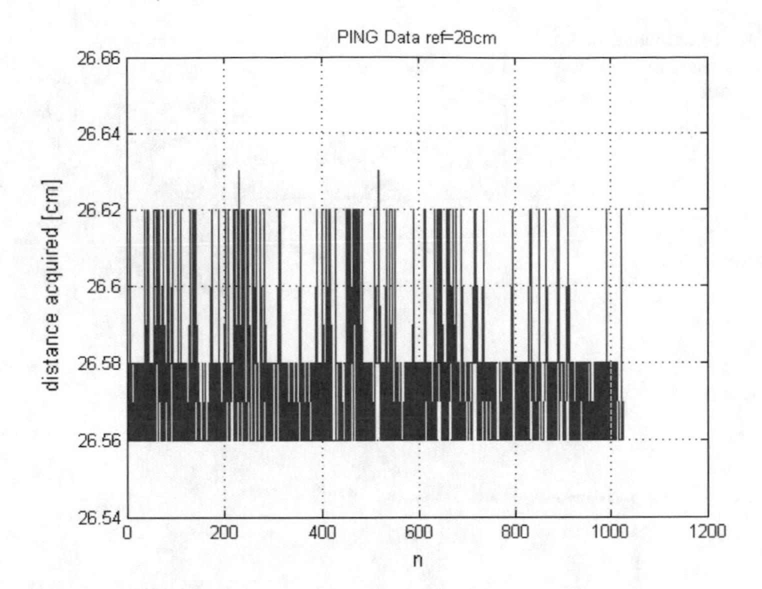

Fig. 3.18 PING))) distance acquisition

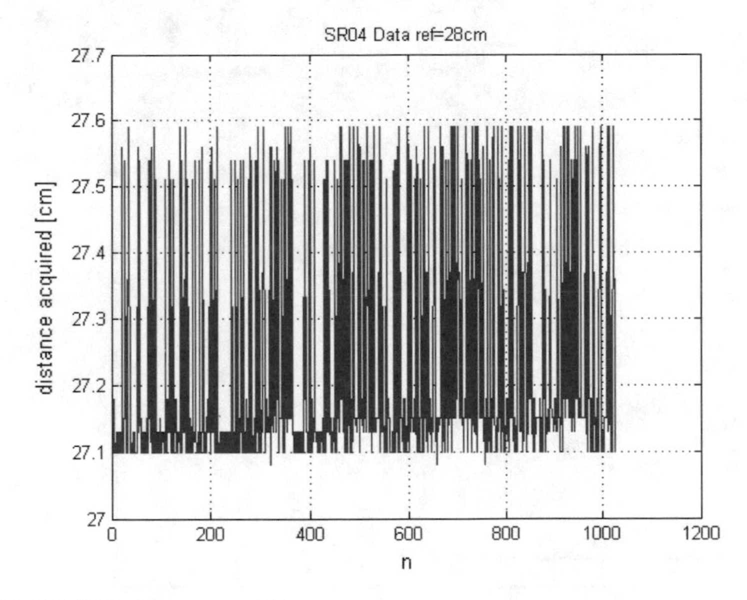

Fig. 3.19 HC-SR04 distance acquisition

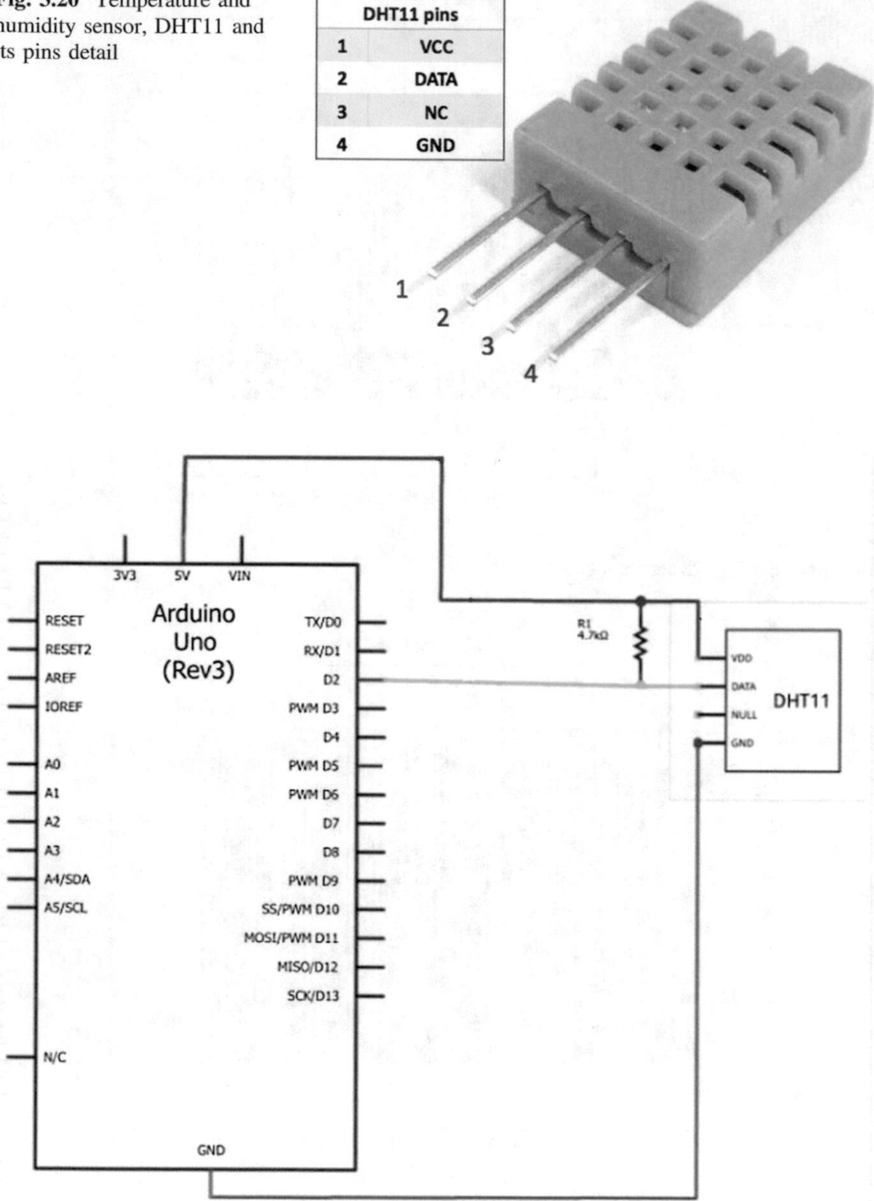

DHT11 pins	
1	VCC
2	DATA
3	NC
4	GND

Fig. 3.20 Temperature and humidity sensor, DHT11 and its pins detail

Fig. 3.21 DHT11/Arduino Uno (Rev3) electric wiring scheme

variable length. If the high level of the pulse is longer than 30 μs then it is a one. If the pulse is less than 26–28 μs then it is a zero (Fig. 3.21).

More details on temperature and humidity extraction from signal are reported in the DHT11 library and in the (DHT11, 2010).

Chapter 4
Integration Among Ultrasonic and Infrared Sensors

4.1 Introduction

In this chapter, another distance sensor was considered in order to integrate the distance measurements for a good UAS navigation information (Sobers et al. 2009; Mustapha et al. 2012). An infrared sensor (IRS) has been considered to extract distance information between UAS and a fixed surface (or obstacle) through infrared (IR) wave bouncing. The IRS distance measurement was useful for a further integration when SRS was out of order or unavailable.

The SRS detects an obstacle like bats or dolphins do, whereas the second one (IRS Sharp) bounces IR off objects to determine how far away they are. The considered distance range is always 20–150 cm, since both sensors work well in this range.

4.2 Infrared Sensor—IRS

The infrared sensor Sharp (GY2Y0A02YK0F) (Sharp 2006) was chosen in this work because it is a miniature, low-cost, fast response time sensor, and has good sensing range to detect obstacles (Fig. 4.1).

The IRS Sharp is a distance measuring sensor unit, composed of an integrated combination of PSD (Position Sensitive Detector), IRED (infrared emitting diode) and signal processing circuit. The variety of the reflectivity of the object, the environmental temperature, and the operating duration are not influenced easily by the distance detection because of adopting the triangulation method. This device outputs the voltage corresponding to the detection distance. Furtherly, this sensor can also be used as a proximity sensor (Sharp 2006) (Table 4.1).

The voltage versus distance curve of the IRS is a nonlinear function (Fig. 4.2). If the output voltage is considered as function of the inverse of distance (1/cm), the

U. Papa, *Embedded Platforms for UAS Landing Path and Obstacle Detection*,
Studies in Systems, Decision and Control 136,
https://doi.org/10.1007/978-3-319-73174-2_4

Fig. 4.1 IRS Sharp (GY2Y0A02YK0F)

Table 4.1 IRS features (Sharp 2006)

Distance measuring range	20–150 cm
Output type	Voltage
Package size (mm)	29.5 × 13 × 21.6
Consumption current (mA)	Typ. 33
Supply voltage (V)	4.5–5.5

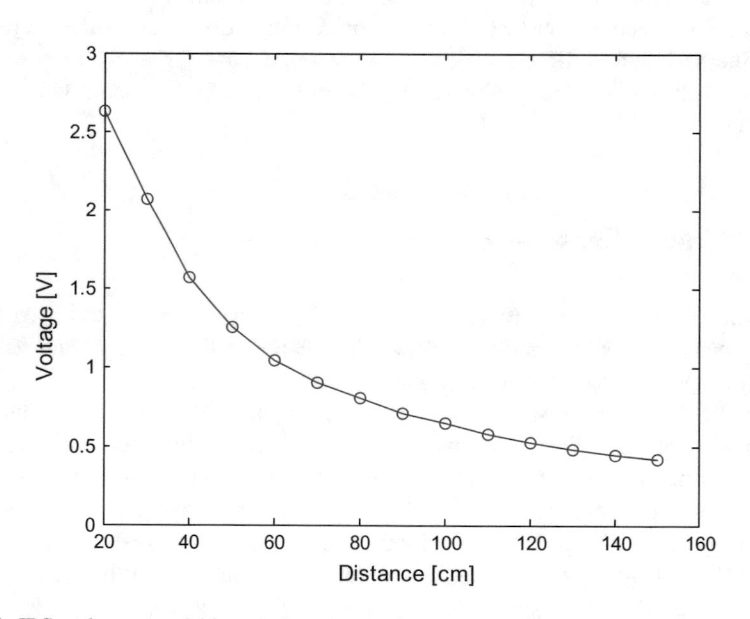

Fig. 4.2 IRS voltage acquisition during data collection, considering a range distance of 20–150 cm

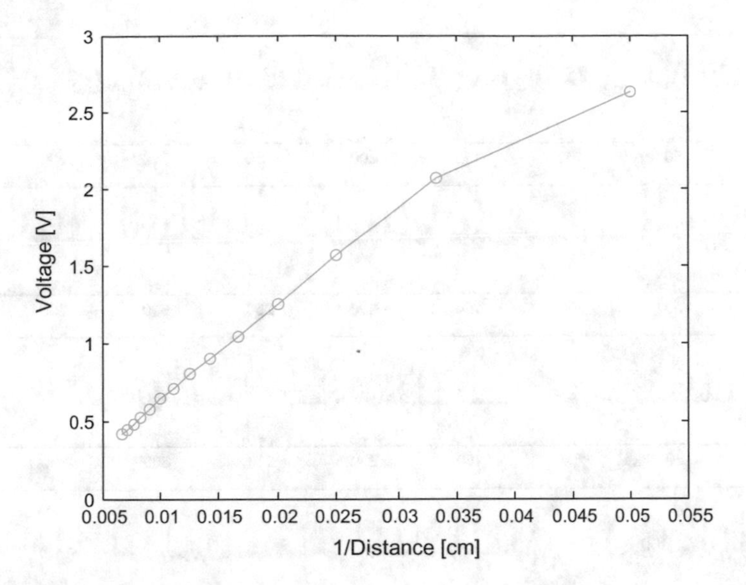

Fig. 4.3 Experimental data collection, (Voltage vs. Reciprocal distance)

characteristic curve becomes piecewise linear over most of the useful sensor range (Fig. 4.3).

Considering the reciprocal distance–voltage relationship, the function became linear over most of the useful range of the sensor. Anyway, log–log scale was an alternative way to linearize and plot nonlinear measurements, but the results were quite similar.

The function in Fig. 4.3 can be split in three linear functions, as shown in the following equations system:

$$\begin{cases} y_1 = 47.689x + 0.3256 \Rightarrow (20-60\,\text{cm}) \\ y_2 = 59.518x + 0.0502 \Rightarrow (60-100\,\text{cm}) \\ y_3 = 69.983x - 0.0573 \Rightarrow (100-150\,\text{cm}) \end{cases} \tag{4.1}$$

where y and x were, respectively, the voltage and the reciprocal distance. In this way, it was possible to estimate distance from the voltage carefully.

An alternative function can be used, considering an exponential formulation (Eq. 4.2), but with less accurate results.

$$y = 61.537x^{-1.048} \tag{4.2}$$

First data collection were made considering a range of distances from 20 to 150 cm, following images shows the voltage data (*y-axis*) versus time (*x-axis*) (Fig. 4.4).

The acquisition time was of 360 s for a fixed distance from 20 to 150 cm. The voltage datum was converted into distance data using the Eq. (4.1) (Fig. 4.5).

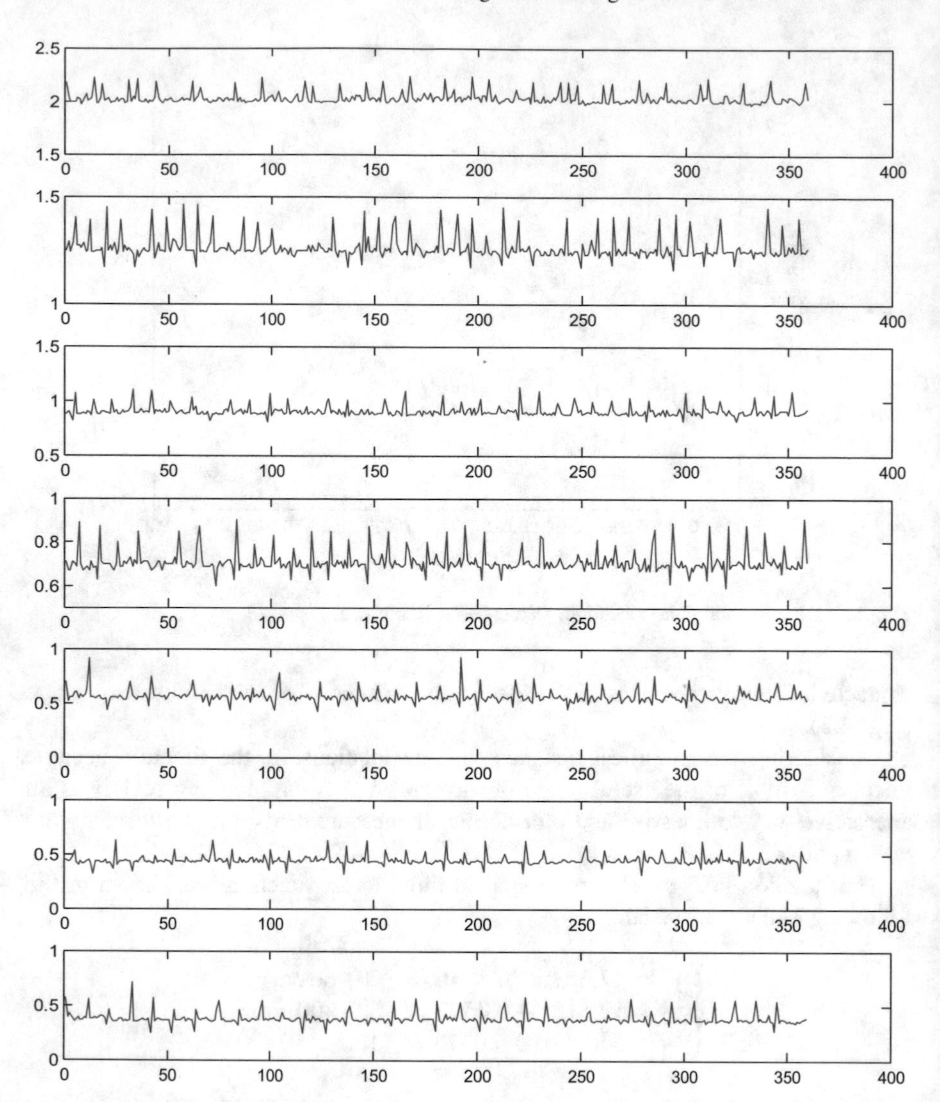

Fig. 4.4 Voltage versus time during IRS distance acquisition. Acquisition time was 6 min (360 s) @ 1 Hz

The IRS was also tested for attitude estimation (Song et al. 2004), in particular during a turn right maneuver (rotation around the roll axes ϕ) of the UAS (Fig. 4.6). The maneuvers were performed at low velocities and low altitude.

The distances extracted from IR sensor, during turn right maneuver, were processed considering the ABC triangle (Fig. 4.6) and the equations for LPTA (landing plane tilt angle—ref. Chap. 2) extrapolation.

The resulting plane was depicted in Fig. 4.7

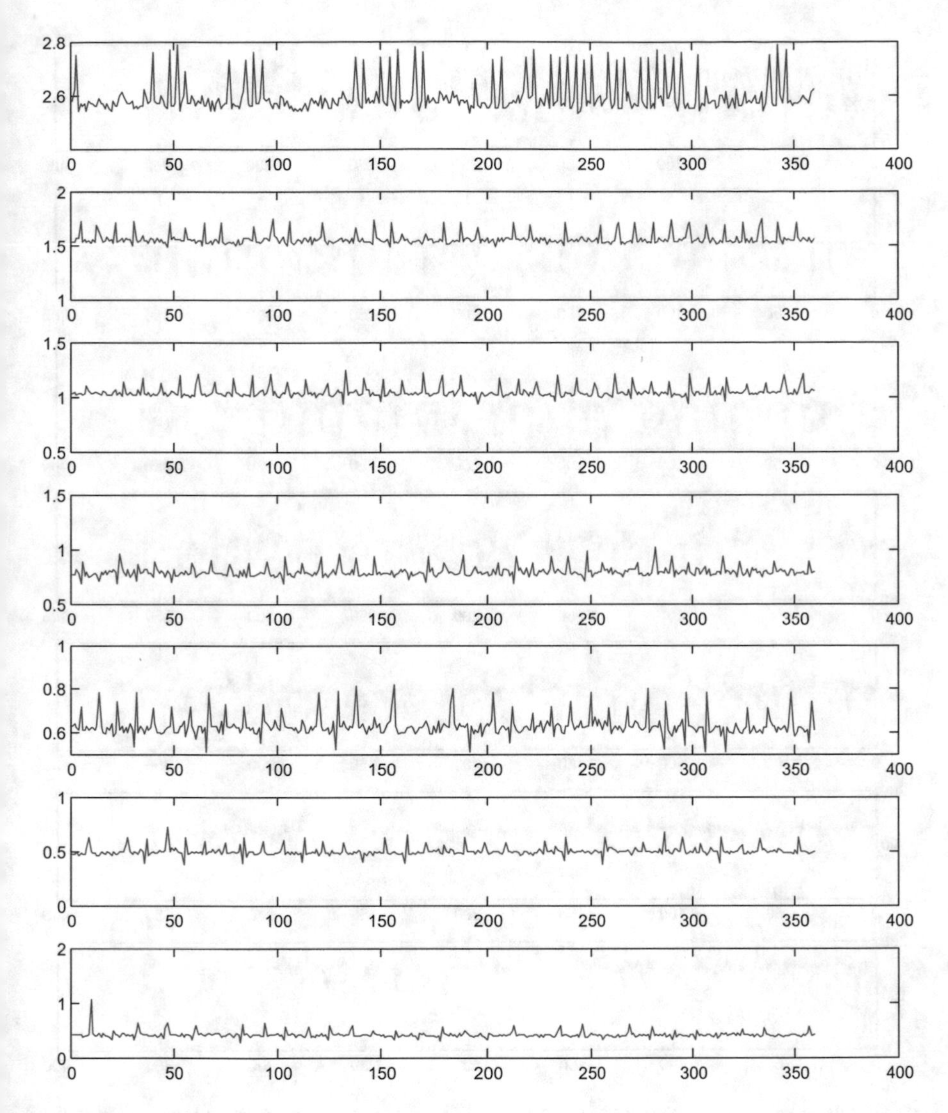

Fig. 4.4 (continued)

The ground reference (blue plane) was inclined and the sensors (UAS) were horizontally aligned.

Figure 4.8 shows the trend of distance acquired; in this case, was considered a distance of 110 cm from the ground.

The distance extraction procedure, of the SRS has been omitted here because it was already considered in Chap. 2.

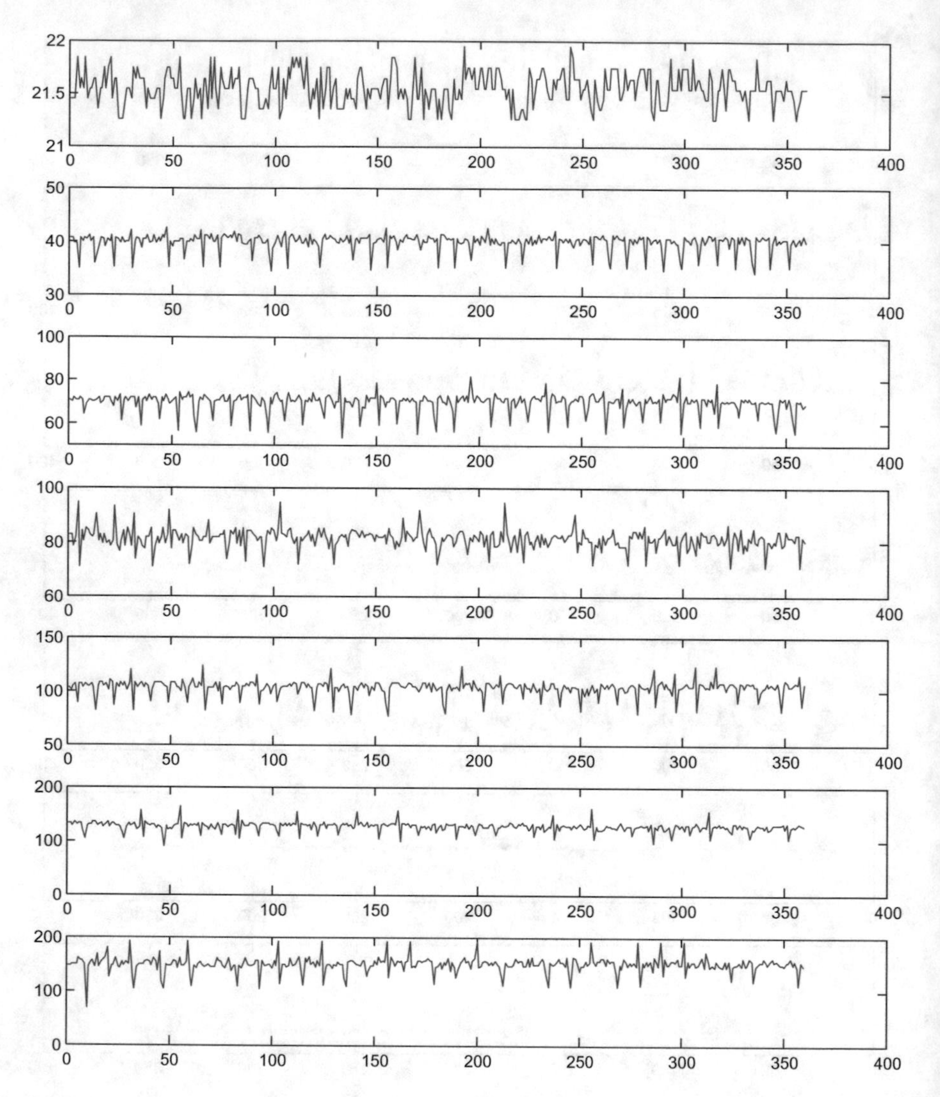

Fig. 4.5 Distance versus time, IRS acquisition. Acquisition time was 6 min (360 s) @ 1 Hz

4.3 Sensors Integration

In the previous section, sensors for distance acquisition have been described. The IRS and SRS data were read, managed, and controlled by an Arduino board, by means of analogical and digital pins. The data were sent from Arduino to PC via serial communication port (or Wireless Shield). The sensor's data were read using pulse-width method (PWM) through an interrupt pin on the Arduino board. The electrical scheme in Fig. 4.9 shows overall distance acquisition system (Timmins 2011).

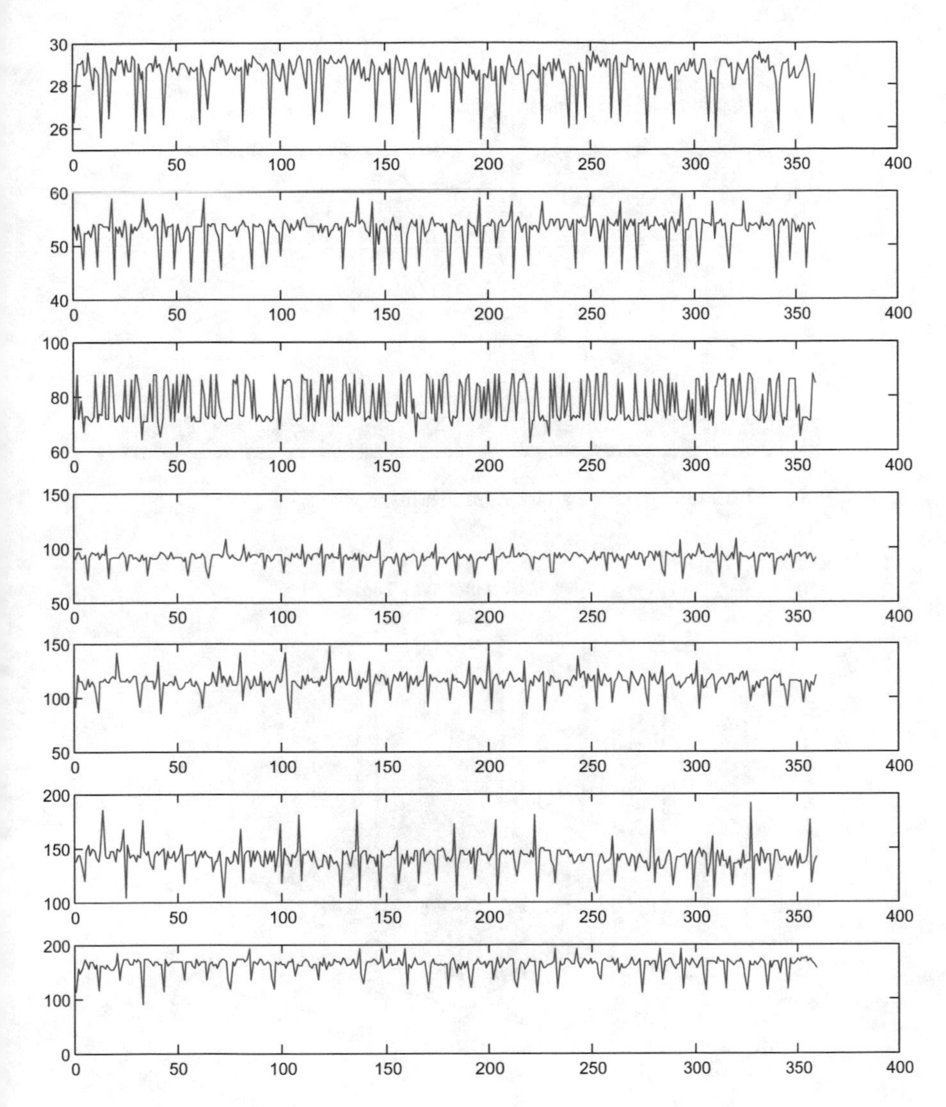

Fig. 4.5 (continued)

The experimental setup acquires and integrates distances from SR and IR sensors, x_{SR} and x_{IR}, with a range of 20–150 cm in 20-cm steps. Each sensor gathers data for 3 min at 2-Hz sampling frequency (one-distance acquisition each 0.5 s). A total of 360 distances per sensor was acquired. The measurement was modeled as

$$\begin{aligned} x_{SR} &= x + e_{SR} \\ x_{IR} &= x + e_{IR} \\ E(e_{SR}) &= E(e_{IR}) = 0 \end{aligned} \tag{4.3}$$

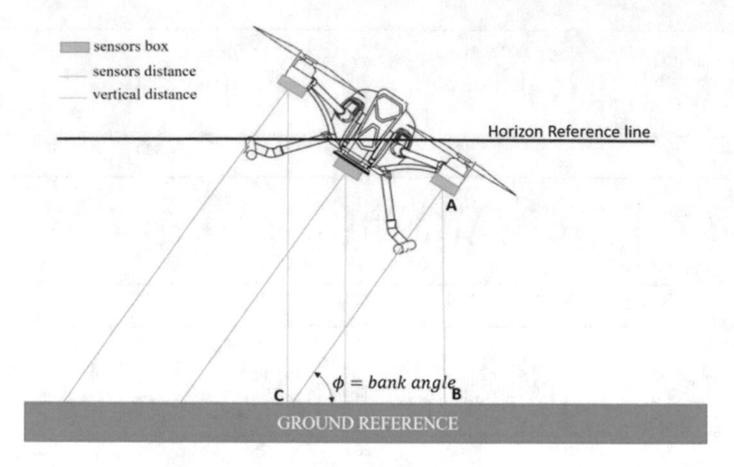

Fig. 4.6 IRS and SRS testing during a turn right maneuver

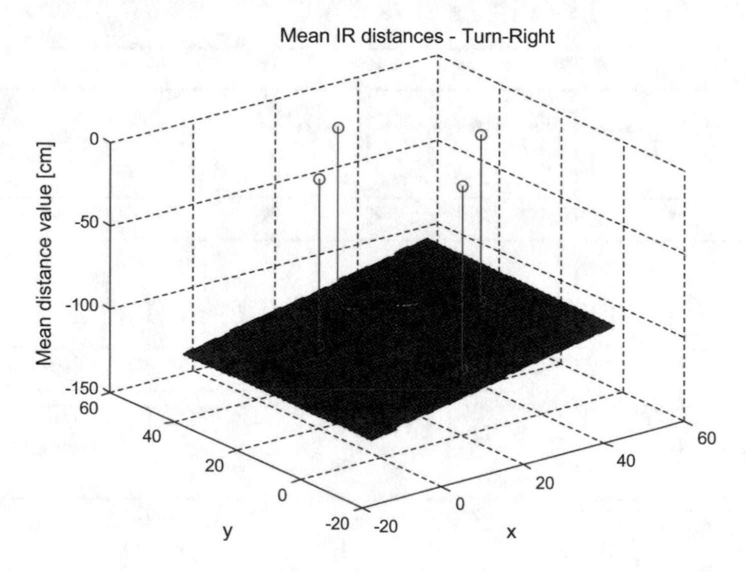

Fig. 4.7 IRS distances and extracted plane during turn right

where x is the true distance, e_{SR} and e_{IR} are the sensor measurement errors, with zero mean (E denotes the expectation) and standard deviations σ_{SR} and $\sigma_{IR,}$ respectively. The simple data fusion technique used in this work combines IRS and SRS measurements in order to obtain an optimal estimate $\hat{x}$ with minimum variance (Gelb 1974). Associating SRS and IRS measurements with weights p_{SR} and $p_{IR,}$ respectively, the mean distance value $\bar{x}$ is

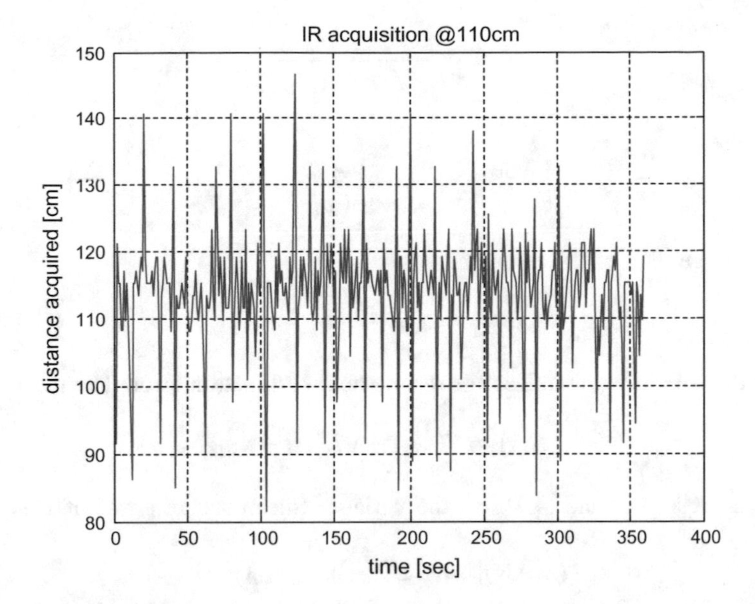

Fig. 4.8 IRS data collection, UAS at 110 cm from the fixed surface (ground), in 360 seconds @ 1Hz.

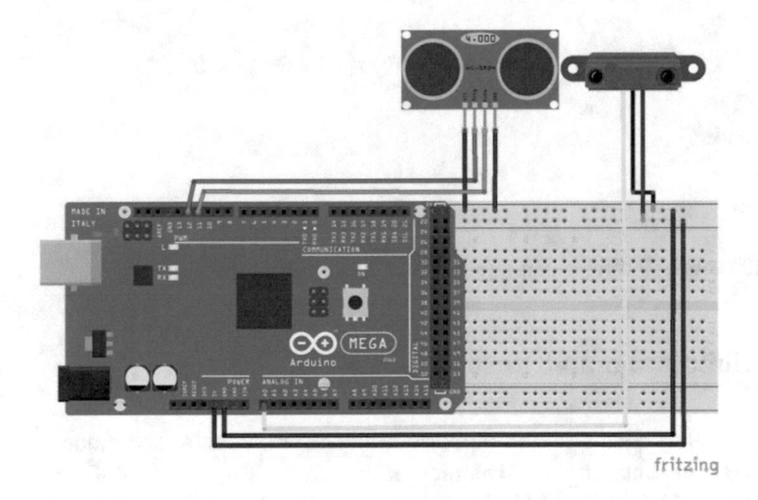

Fig. 4.9 IRS and SRS acquisition system and wiring with Fritzing software

$$\bar{x} = \frac{p_{SR}x_{SR} + p_{IR}x_{IR}}{p_{SR} + p_{IR}} \tag{4.4}$$

Letting

$$w = \frac{p_{IR}}{p_{SR} + p_{IR}} \tag{4.5}$$

The expression of the estimator $\hat{x}$ is

$$\hat{x} = \bar{x} = x_{SR} + w(x_{IR} - x_{SR}) \tag{4.6}$$

where the weight w minimizes the mean square estimation error. The mean of $\hat{x}$ is

$$E(\bar{x}) = E(x_{SR}) + wE(x_{IR} - x_{SR}) \tag{4.7}$$

Using Eqs. (4.6) and (4.7), σ^2, the variance (mean square error) of $\hat{x}$, is

$$\begin{aligned} \sigma^2 &= \{(1-w)[x_{SR} - E(x_{SR})] + [wx_{IR} - E(x_{IR})]\}^2 \\ \sigma^2 &= (1-w^2)\sigma_{SR}^2 + w^2\sigma_{IR}^2 \end{aligned} \tag{4.8}$$

Variance minimization implies that

$$\frac{\partial \sigma^2}{\partial w} = -2(1-\hat{w})\sigma_{SR}^2 + 2\hat{w}\sigma_{IR}^2 = 0 \tag{4.9}$$

Whence

$$\hat{w} = \frac{\sigma_{SR}^2}{\sigma_{SR}^2 + \sigma_{IR}^2} \tag{4.10}$$

and the corresponding mean square estimation error, $E(\hat{x}^2)$, is $\sigma_{SR}^2\sigma_{IR}^2/(\sigma_{SR}^2 + \sigma_{IR}^2)$.

4.4 Simulation and Results

The system has been tested in indoor/outdoor environments. The indoor tests were performed in our Flight Dynamics Laboratory (Department of Science and Technology, University of Naples Parthenope, Italy), installing on a test bench the distance measurement platform and acquiring distance data with the sensors installed on board the UAS. Two scenarios were devised for the experimental setup

- Measurements in International Standard Atmosphere (ISA: $T = 15$ °C, pressure = 1013 mbar, dry air);
- Measurements in ISA+20 (worst case).

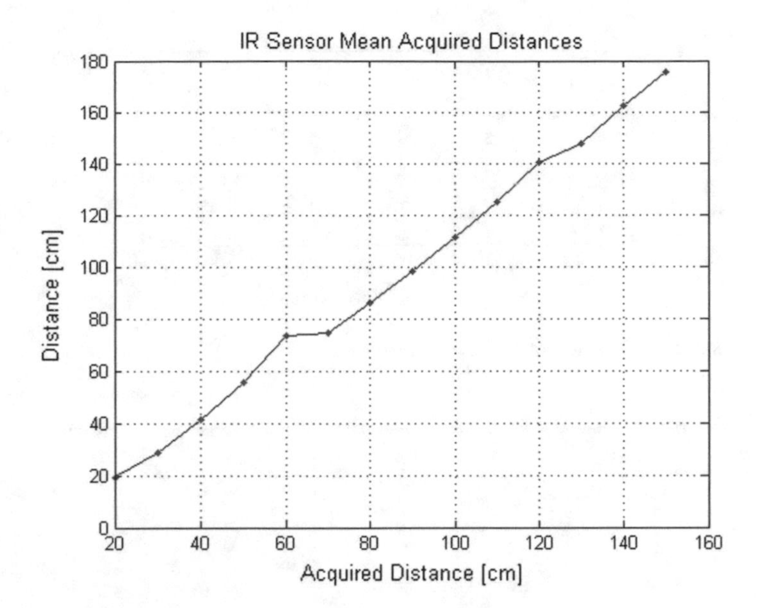

Fig. 4.10 Measurement results for the IR sensor

The optimal distance estimator and the weight are given by Eqs. (4.6) and (4.10) of the preceding section.

The type of material used as the obstacle is a white paper box. Other data session considered boxes of different colors, but the results were quite similar, as noted in (Mustapha et al. 2012). The voltage–distance characteristic from 20 to 150 cm, obtained from IR measurements, interpolated with Eq. (4.1), is shown in Fig. 4.10, whereas Fig. 4.11 shows the standard deviation of x_{IR} as a function of the distance, and it increases if the acquired distance increases.

Figure 4.12 shows the experimental results for the distance measurements of the SR sensor, plotting the standard deviation σ as a function of the measured distance (Fig. 4.13).

If the obstacle is closer, the error of both sensors decreases. In ISA conditions, the SR sensor seems to behave better than the IRS. Combining the two sets of measurements using Eqs. (4.6) and (4.10), the resulting estimated distance is depicted in Fig. 4.14, and standard deviation in Fig. 4.15.

Table 4.2 shows *rms* (root mean square) error and max distance error for IRS data, SRS data and combined data.

As far as the worst-case data are concerned, the environment temperature increased from 15 to 35 °C. We noted that the individual IRS and SRS distance measurements errors were significantly worse than those evaluated in the ISA scenario, whereas the optimal estimation (IR and SR data fusion) has a still acceptable error (Figs. 4.16 and 4.17). For comparison with the measurements in the ISA scenario, Table 4.3 compares *rms* and maximum distance errors of IRS, SRS, and combined data in the ISA+20 scenario.

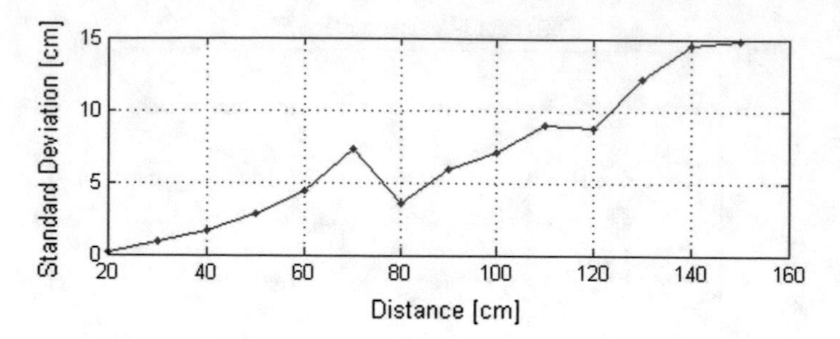

Fig. 4.11 Standard deviation of IRS measurements

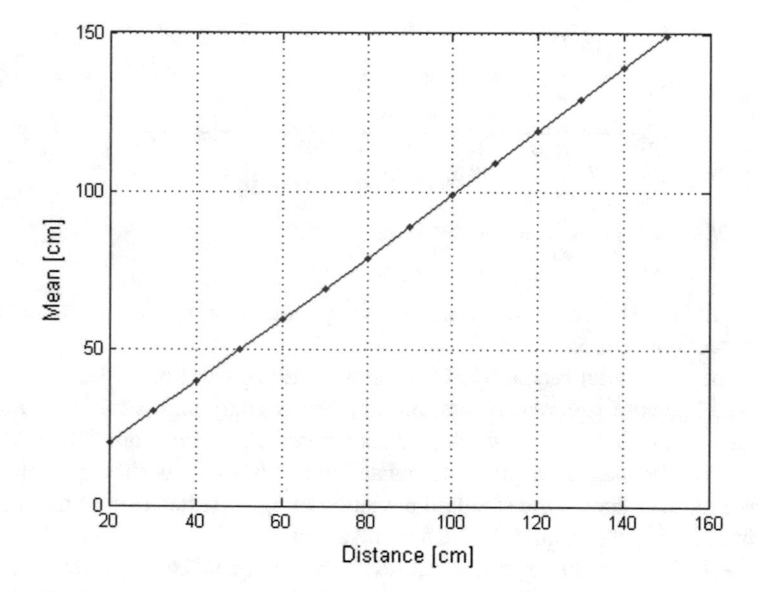

Fig. 4.12 Measurement results for the SR sensor

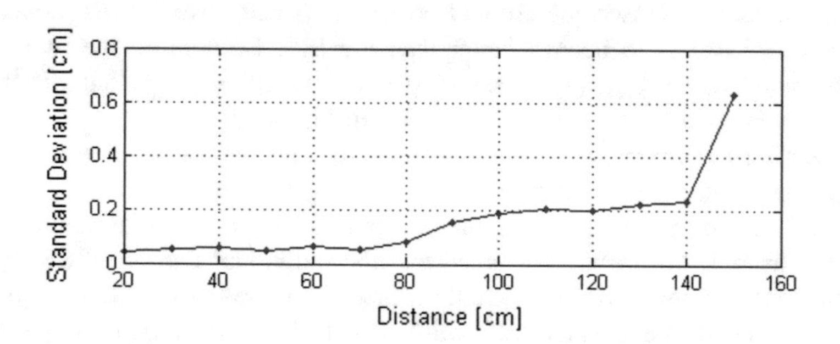

Fig. 4.13 Standard deviation of SRS measurements

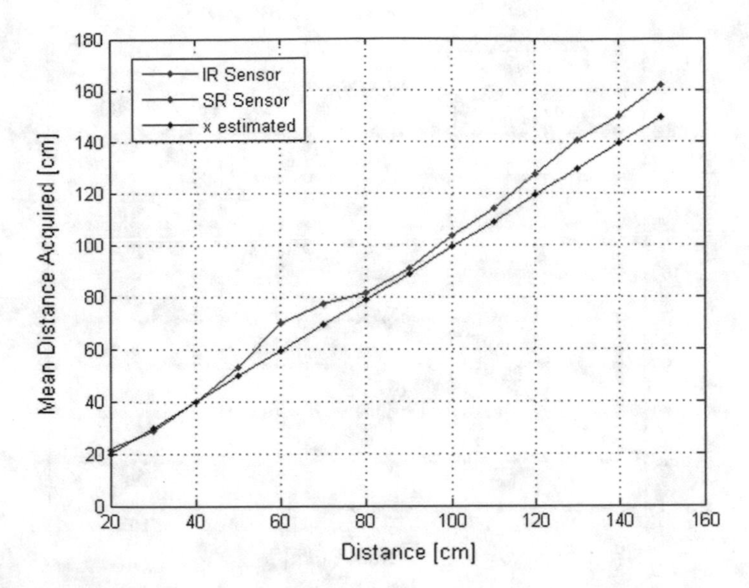

Fig. 4.14 Data fusion: combined estimated distance versus individual measurements (IRS and SRS), ISA scenario

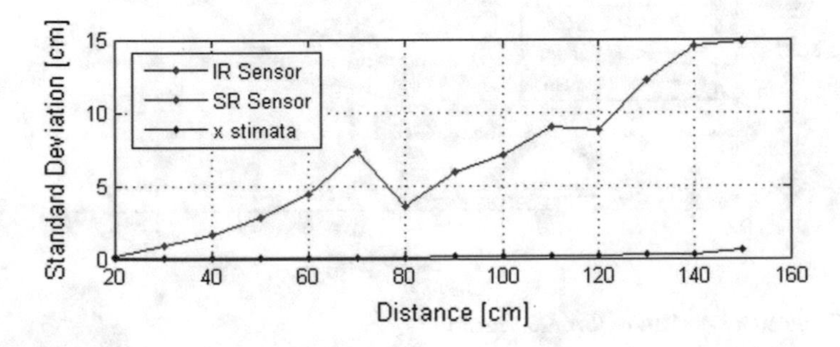

Fig. 4.15 Standard deviations (ISA scenario)

Table 4.2 Error comparison (IRS, SRS, and combined data), ISA scenario

	IRS data	SRS data	$\hat{x}$
rms error (cm)	6.4325	0.8489	0.8248
maxErr (cm)	11.705	0.192	0.116

The optimal distance estimator exhibits 92% increment on the *rms* in the ISA+20 scenario, which is still satisfactory for a safe landing maneuver of the quadrotor. Use of only SRS data for distance estimation would have given an *rms* error approximately double with respect to the combined usage of IRS and SRS data.

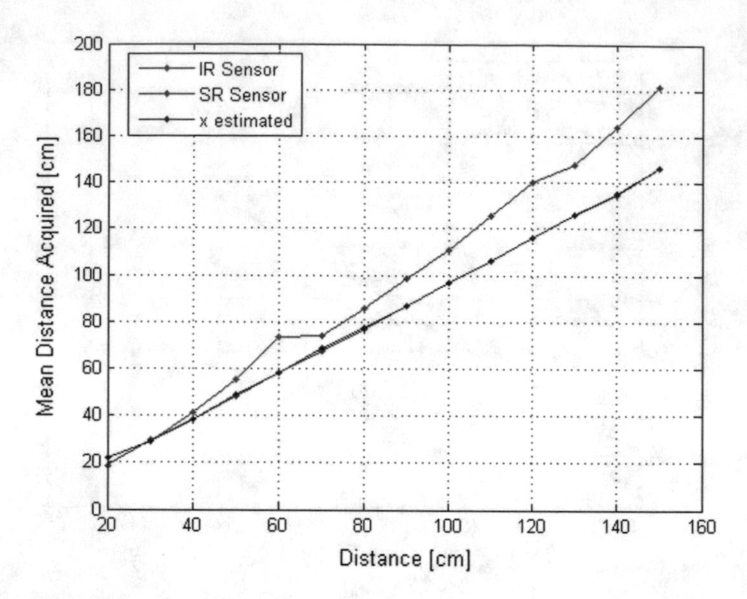

Fig. 4.16 Distance measurements (ISA+20 scenario)

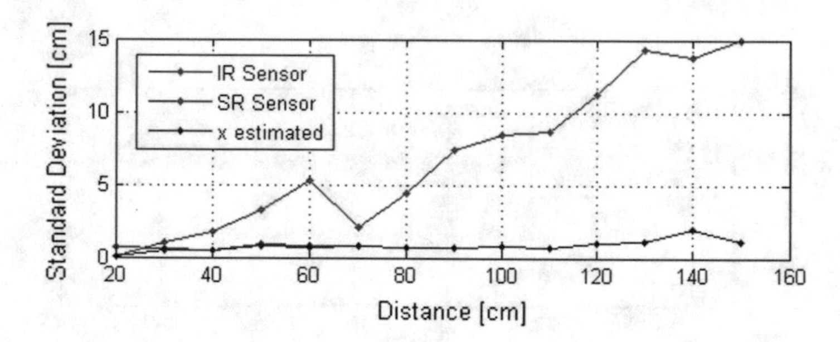

Fig. 4.17 Standard deviation, ISA+20 scenario

Table 4.3 Error comparison, ISA+20 scenario

	IRS data	SRS data	$\hat{x}$
rms error(cm)	14.643	3.093	1.587
maxErr (cm)	21.437	0.6089	0.1910

4.5 Conclusion

This paragraph presents a simple minimum square error estimation technique that combines distances from the ground or from an obstacle measured by low-cost and lightweight sonic sensor and an infrared sensor on board a small commercial quadrotor. This is the first step towards reliable automatic or aided landing procedures and a full "sense-and-avoid" (obstacle detection) and attitude estimation system. Test results, performed by measuring distances from various kinds of obstacles and ground surfaces show that the data fusion technique allows us a better estimation, by weighting the observations with coefficients depending on the sensor measurement errors. Different scenarios were simulated, and in worst-case conditions (ISA+20) the *rms* estimation error remained significantly lower than the single sensor errors (about 1.6 cm against 3.1 cm for SRS data and 15 com for the IRS measurements). In standard (ISA) operating conditions, the use of the sole SR sensor could be sufficient (see Tables 4.2 and 4.3). The degradation of the sensors' measurement performances has been found to be conveniently mitigated by the data fusion technique.

Additional sensors (laser or optical) and of more complex estimation techniques (Kalman filtering, numerical fuzzy integration) can be useful in order to have more robust estimates of the distance from obstacles in the mission area, of the distance from ground during a landing procedure, and of UAS attitude. Implementing an effective automatic methodology for sense-and-avoid algorithms on board the UAS, providing autonomous navigation capabilities, is the long-term objective of this book.

Chapter 5
Optical Sensor for UAS Aided Landing

5.1 Introduction

The installation of an optic sensor (i.e., camera) to a UAS allows the vehicle to perform a variety of tasks autonomously (Valavanis 2007). This chapter presents UAS vision system developed and tested at the Parthenope University of Naples (Department of Science and Technology) to perform an aid for the UAS during the landing procedure at low altitude.

Visual sensors, such as electro-optic or infrared cameras, have historically been included in the onboard equipment of many UASs to increase their value as a surveillance tool.

In this chapter, an autonomous vision-based landing system for small quadrotor was designed and developed (Sharp et al. 2001). The system uses a single camera to determine the precise position and orientation respect a well-defined landing pattern. Indeed, the developed procedure is based on photogrammetric space resection solution (SRS), which allows determining a single camera position and orientation starting from at least three reference control points, not aligned, whose image coordinates may be measured in the image camera frame. Obviously, the computation of 3D position and attitude parameters of UAS is carried out in the reference system of control points, and it can easily be expressed in a global reference system.

A specific landing pattern with five circular colored targets was realized; therefore for each image, the 2D image frame coordinates of the target center were extracted through a particular algorithm. The center and radius of the circular colored target were found by using Hough Transform Function algorithms.

The aim of this section is to compute UAS precise position, from single image, in order to have a good approach to landing field. This procedure is helpful when the GPS module is out of order or a sufficient number of satellites are not available; otherwise, GPS tracking can be used for landing position correction. Furthermore, this procedure can be applied when the landing field is movable; the UAS will follow the landing pattern until the landing phase is closed.

U. Papa, *Embedded Platforms for UAS Landing Path and Obstacle Detection*,
Studies in Systems, Decision and Control 136,
https://doi.org/10.1007/978-3-319-73174-2_5

5.2 Vision-Based Embedded System

This section explains how to search location and orientation of onboard camera, especially for a quadcopter UAS, through a vision-based methodology. In particular, the space resection method (SRM) used in photogrammetry is adopted. The approach performs a set of landing path coordinate corrections, in order to have an accurate path of descent in a specific area, aided by a landing pattern on the ground positioned. Thus, the UAS receives from the system a track correction (*x*, *y*, *z*) for overall landing procedure. The corrections are sent to the autopilot system, but this is not dealt in this work. Finally, it is important to specify that the overall system be embedded on board, and on the ground, there is only a landing path that contains a specific coded target.

The whole system is composed of the following elements:

- Camera module—Raspberry Pi camera module;
- Minicomputer—Raspberry Pi mod. B.

5.2.1 Camera

The camera module used in this project is RPI CAMERA BOARD, i.e., Raspberry Pi camera module as shown in Fig. 5.1 (Raspberry Pi Camera 2016). The camera plugs directly into the CSI (camera serial interface) connector placed on the Raspberry Pi (Fig. 5.2). The CSI bus is capable of extremely high data rates, and it exclusively carries pixel data to the BCM2835 processor.

In terms of resolution, this module delivers clear 5MP resolution (2592 × 1944 pixels) image through a Omnivision 5647 sensor in a fixed-focus camera that supports HD 1080p30, 720p60, and VGA90 video modes, recording, respectively, at 30, 60, and 90 frames per second. The video mode will be suitable when the video system will replace the photos with the video. The Pi camera can be accessed through the MMAL and V4L APIs, and there are numerous third-party libraries built for it, including the Picamera Python library (Fig. 5.3).

In addition, a dongle Wi-Fi antenna is installed on the Raspberry, in order to receive the data (photos, video, etc.) in real time and check all the phases of the visual landing operation (Fig. 5.4).

The visual system is powered by a battery bank linked to the Raspberry via micro-USB port. In particular, the battery used for the first survey is a 9 V battery.

For evaluating the current vertical coordinate, also barometric height measurement @MSL was used; alternatively, it is possible to use an Ultrasonic Sonar Sensor designed in previous works (Eisenbeiss 2004). For completeness of information, the barometric pressure and sonar sensor were respectively the BMP 180 and HC-SR05. Furthermore, it is possible to add a GPS module to extract correct altitude and position of the UAS for comparing and merging data.

Fig. 5.1 Raspberry camera board (5MP, 1080p, v1.3)

Fig. 5.2 Raspberry Pi2

On this system, OpenCV libraries are loaded that allow extracting target contour, and finally, coordinates in the current image are acquired. Basic test landing field, shown in Fig. 5.5, is used.

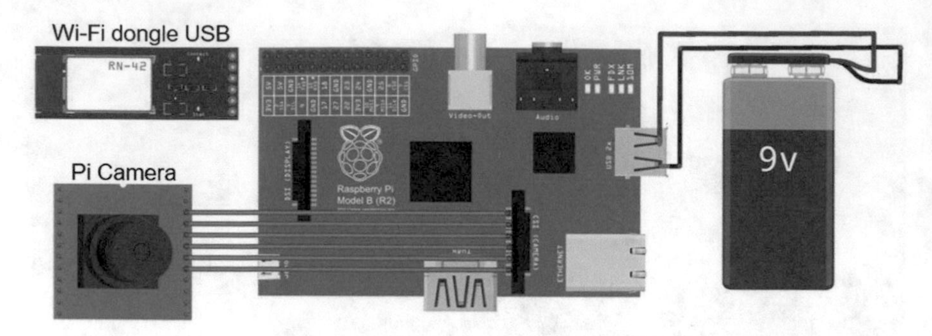

Fig. 5.3 Electrical scheme of the vision system installed on the UAS

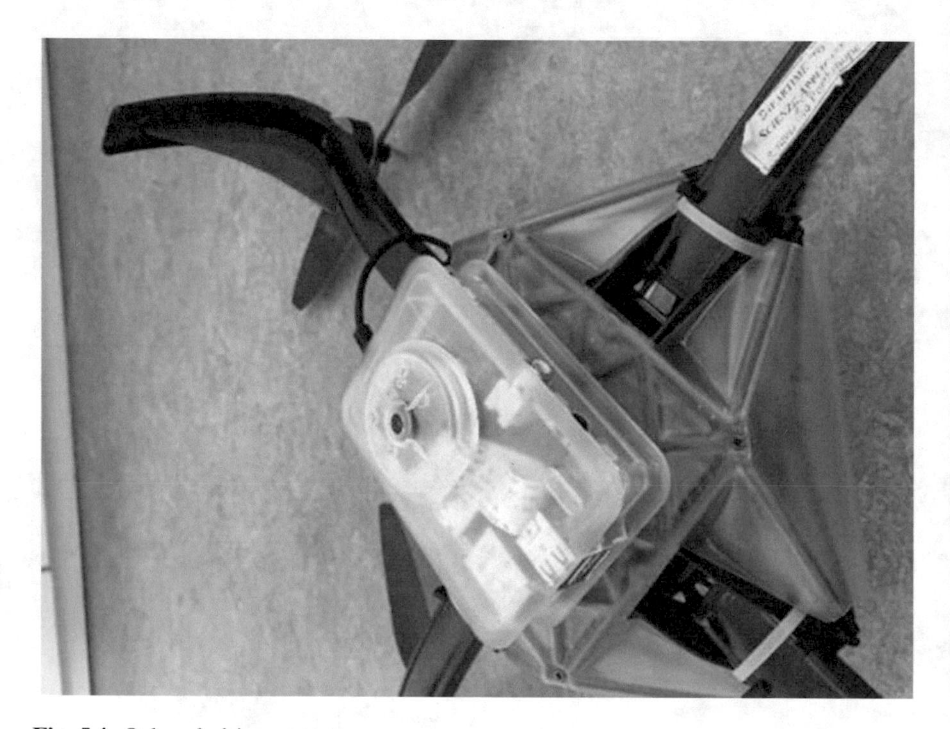

Fig. 5.4 Onboard vision system

The software is able to recognize the circular target, filtered by color, and extract its coordinate (in pixel). This is possible trough Circular Hough Transform function. These functions are developed due to OpenCV libraries. Input parameter for good image circle finding is radius range into images acquired; it depends on the range of altitudes of UAS (in this case 20–150 cm).

The coordinates extracted in real time are depicted in Fig. 5.6.

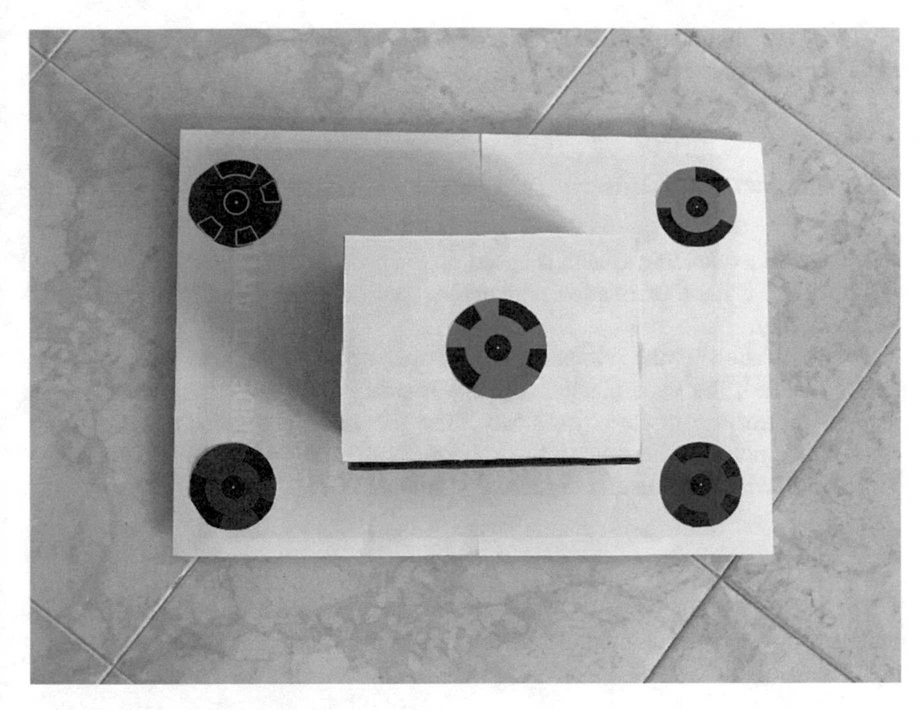

Fig. 5.5 Basic test landing field with five color coded target

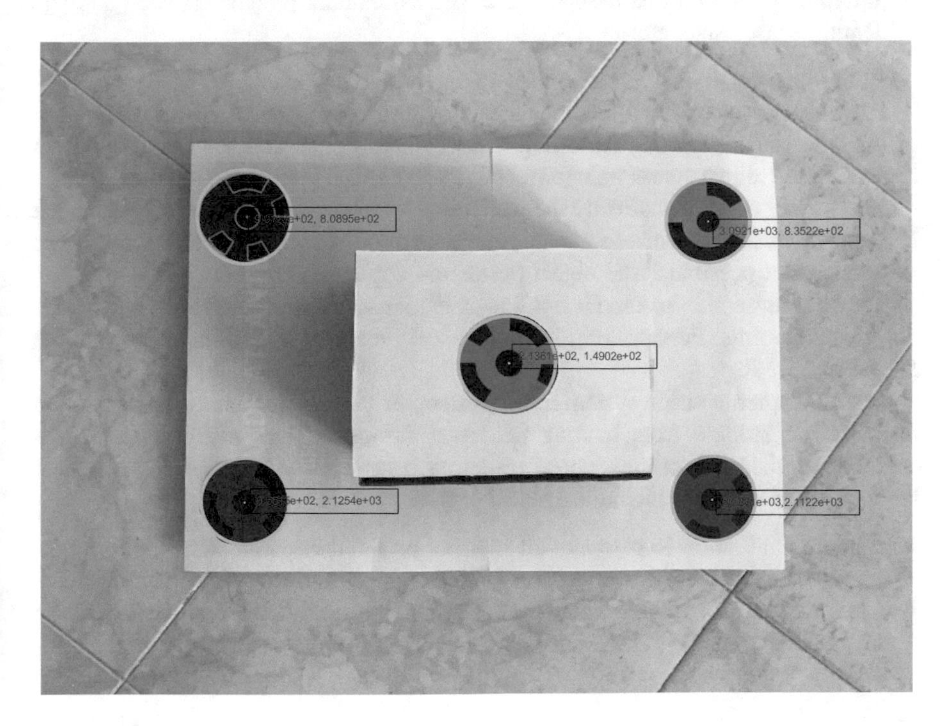

Fig. 5.6 Target reconnaissance on pattern

A range of altitude from 20 to 150 cm is chosen, because this range is typical for a characterization of the UAS landing procedure.

5.3 Extract the Camera Position and Attitude

The vision-based landing system is based on the processing of a single image. Goal was to detect camera orientation parameters (position and attitude) in a fixed reference system.

The realization of the reference system was obtained from a special landing pattern. Indeed, the determination of full orientation assumes the availability of enough information in the object space. The mapping of a point x in the object space into a point in the camera space (expressed in homogenous coordinates) is fully described by a projection matrix P (Luhman 2011):

$$x' = Px \tag{5.1}$$

The projection matrix is a 3×4 matrix that describes the orientation of a pinhole camera. In photogrammetry, the orientation is divided into two different types:

- External orientation that describes the position and the attitude of a camera.
- Internal orientation that describes the internal camera parameters such as focal length, sensor size, etc.

Both these parameters are absorbed in the projection matrix.

In photogrammetry, two methods are known to compute the correct orientation of a single camera starting from object coordinate points and the homologous in camera space: direct linear transform (DLT) and space resection (SRS).

The former computes directly the projection matrix, solving a linear system, but it needs at least six corresponding image and object points. Furthermore, the solution is not possible if the object points are coplanar (Faugeras 1993).

The space resection method is not linear, but it assures a solution with only three, not aligned, points. Furthermore, the theoretical precision of SRS is better than the DLT.

In this chapter, a vision system for a quadcopter UAS which estimates its relative position and attitude from landing field was designed. This vision system uses customized vision algorithms space resection based. All procedures for a correct landing occur in the following stages:

- Camera calibration to compute all internal orientation parameters;
- Image coordinates point extraction (image space); and
- Computation of external orientation parameter extrapolation (orientation phase using SRS).

Interior orientation parameters are extracted from calibration procedure of the camera. Exterior orientation parameter is extracted by means of space resection method.

Next section explains overall system setup and initial procedure in order to make a correct data extraction.

5.3.1 *Camera Calibration*

To obtain external orientation parameters with good accuracy, it is mandatory to calibrate the camera. The calibration is the process that allows determining the internal orientation of a camera and the distortion coefficients.

In particular:

- Principal point or image center (x_0', y_0');
- Focal length (f); and
- Radial distortion coefficient (K_1, K_2).

Brown originally developed one of the most used analytical camera calibration techniques. This method is often used in close-range photogrammetry to obtain the internal parameters with high accuracies.

Camera raspberry camera module (Fig. 5.1) is able to fix the focal length and to acquire a photo every second on a full resolution (about 5 Megapixel). To avoid overloading the buffer and the computation and to obtain images with low brightness, the resolution was set to 1920 × 1080 (about 2 Megapixel), according to an HD video.

Camera calibration was carried out using a set of coded circular target, in order to obtain subpixel precision and to automatize the procedure of calibration (Fig. 5.7).

A guided procedure, using specific software (e.g., Photomodeler), was done, where several frames were acquired for the calibration procedure; in detail, eight images were acquired. They should depict the test field perpendicularly and obliquely, and each image should have a relative rotation of 90° around the optical axis (Table 5.1).

The calibration procedure was done, giving the following results for the Pi camera.

5.3.2 *Point's Extraction*

The object recognition is an important task in computer vision; it identifies a specific object on the image. Both SRS and DLT need to identify the ground control point position on the image frame.

In order to solve this problem, a specific landing pattern, employing circular target, was developed (see next section). Circular targets were designed both to facilitate automatic detection and to assure a good marking accuracy; indeed, it is

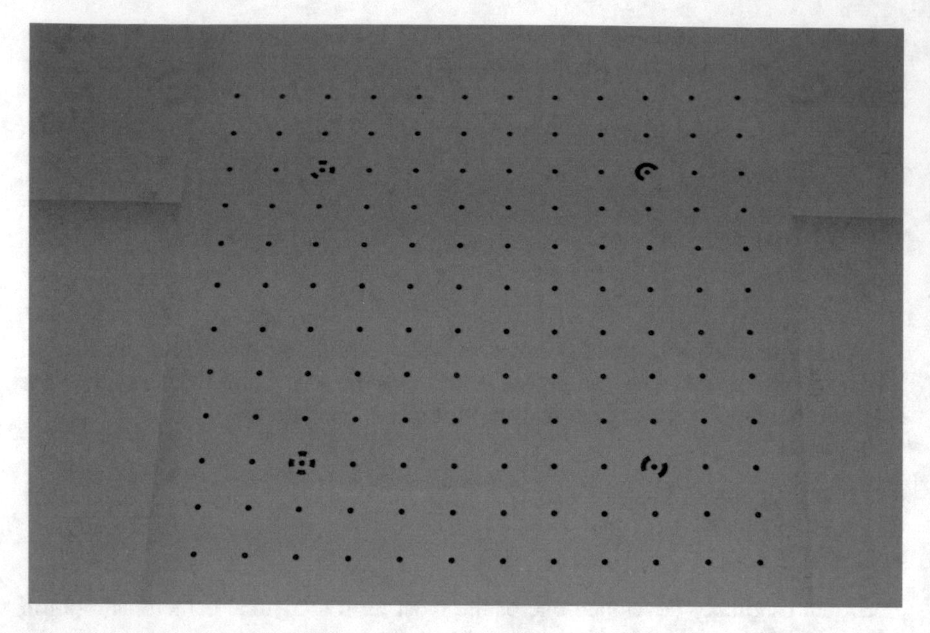

Fig. 5.7 Pattern used for camera calibration procedure

Table 5.1. Internal orientation parameters for Pi camera obtained, thanks to Photomodeler

Focal length (mm)	5.94	3.0E-03
Sensors size (mm)	6.0 (wide) 3.38 (high)	6.0E-03 5.0E-05
Principal point	X: 3.001 Y: 1.703	5.5E-14
Lens distortion radial coefficients	K_1: −1.932E-03 K_2: 1.553E-04 K_3–K_6 = 0.0	1E-04 1.9E-05

well known that subpixel accuracy can be achieved. All the measurements are referred to the center of circular target.

Several methods exist for the automatic detection of circular targets in an image. They can be divided into two categories:

- No-initial approximation required methods: these approaches allow to obtain coordinates center of a circular target on image without any initial approximation;
- Initial approximation required methods: these approaches start from an approximated position of the target center, and they are able to detect the center with high accuracy.

Of course, the target coordinate center is not sufficient to detect the correspondent GCP. The matching is performed using a unique code for every target. In this work, the codification is based on the color target.

5.3.3 *Orientation*

The camera orientation is computed by resection with respect to a landing frame. The resection is a linearization of collinearity Eq. (5.2), to describe the transformation of object coordinates (X, Y, Z) into corresponding image coordinates (x', y').

$$\begin{aligned} x' &= x_0 + c\frac{r_{11}(X - X_0) + r_{12}(Y - Y_0) + r_{13}(Z - Z_0)}{r_{13}(X - X_0) + r_{32}(Y - Y_0) + r_{33}(Z - Z_0)} + \Delta x' \\ y' &= y_0 + c\frac{r_{21}(X - X_0) + r_{22}(Y - Y_0) + r_{23}(Z - Z_0)}{r_{13}(X - X_0) + r_{32}(Y - Y_0) + r_{33}(Z - Z_0)} + \Delta y' \end{aligned} \tag{5.2}$$

where $r_{i,j}$ are elements of a 3D rotation matrix R:

$$\begin{bmatrix} x' \\ y' \\ c \end{bmatrix} = \begin{bmatrix} X - X_0 \\ Y - Y_0 \\ Z - Z_0 \end{bmatrix} \tag{5.3}$$

The image coordinates are function of the interior orientation parameters: x_O, y_O, c, $\Delta x'$, and $\Delta y'$ are, respectively, principal point coordinates, camera focal length, and relative deviation due to distortion effects and exterior orientation parameters (X_O, Y_O, Z_O, ω, φ, κ). The first three are the positions of the perspective center in object reference system, while the others are rotation angles: ω (tilt horizontal axis), φ (roll around azimuth axis), and κ (roll around optical axes), as shown in Fig. 5.8.

The collinearity equation system (5.2) can be rewritten in the following system of correction equations:

$$\begin{aligned} x' &= x'_0 + vx' \\ y' &= y'_0 + vy' \end{aligned} \tag{5.4}$$

where x_0', y_0' are the image coordinates computed using approximated orientation parameters, while vx' and vy' are the adjustment parameters to obtain the correct x' and y' image coordinates. The adjustment parameters can be obtained as linearization of Eq. (5.2), using as initial point the approximated orientation parameters employed to compute x_0', y_0':

$$\begin{aligned} vx' &= \left(\frac{\partial x'}{\partial X_0}\right)_0 \mathrm{d}X_0 + \left(\frac{\partial x'}{\partial Y_0}\right)_0 \mathrm{d}Y_0 + \left(\frac{\partial x'}{\partial Z_0}\right)_0 \mathrm{d}Z_0 + \left(\frac{\partial x'}{\partial \omega}\right)_0 \mathrm{d}\omega + \left(\frac{\partial x'}{\partial \varphi}\right)_0 \mathrm{d}\varphi + \left(\frac{\partial x'}{\partial \kappa}\right)_0 \mathrm{d}\kappa \\ vy' &= \left(\frac{\partial y'}{\partial X_0}\right)_0 \mathrm{d}X_0 + \left(\frac{\partial y'}{\partial Y_0}\right)_0 \mathrm{d}Y_0 + \left(\frac{\partial y'}{\partial Z_0}\right)_0 \mathrm{d}Z_0 + \left(\frac{\partial y'}{\partial \omega}\right)_0 \mathrm{d}\omega + \left(\frac{\partial y'}{\partial \varphi}\right)_0 \mathrm{d}\varphi + \left(\frac{\partial y'}{\partial \kappa}\right)_0 \mathrm{d}\kappa \end{aligned} \tag{5.5}$$

Generally, the approximated position of the camera does not provide good results, although they are corrected using a linearization procedure. Indeed, the distance between the computed coordinates, using the abovementioned linearization

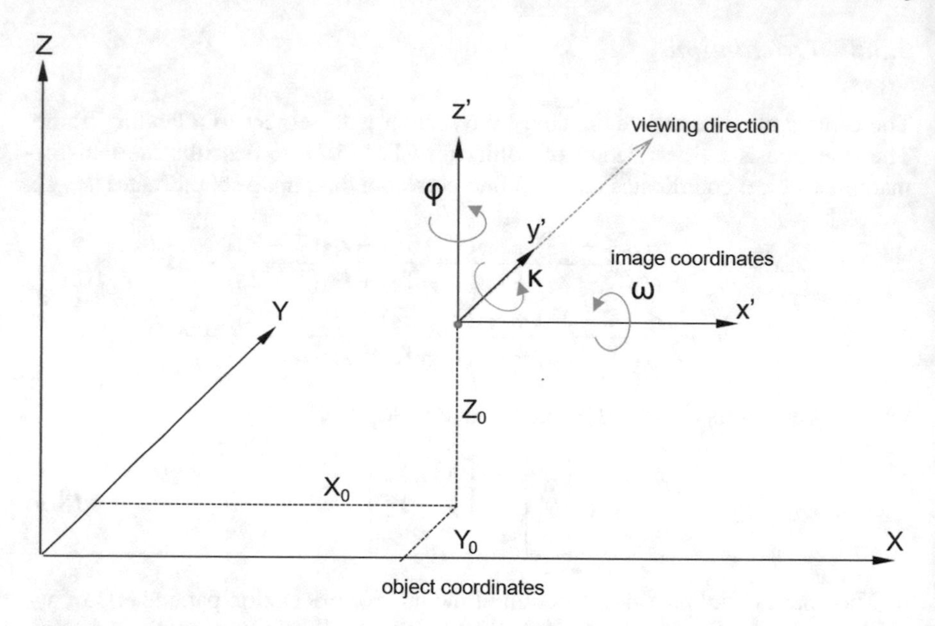

Fig. 5.8 Two reference systems: the object (XYZ) and the camera one (x' y' z')

procedure, and the measured coordinates approaches zero after several iterations only. At the end of the iterative process, the method is able to estimate the camera attitude and position (Fraser 1997).

In order to solve the system in Eq. (5.4), we need at least three points (not aligned); indeed, such observations provide six equations, which allow estimating the six external orientation parameters. Of course, to increase the redundancy of the equation system (5.4), it needs to add further observations, and the solution can be estimated using the classical least-square adjustment method.

5.4 Validation Procedure

In order to inspect the accuracy and precision achievable employing the proposed methodology, a validation procedure was designed and then performed.

The aim of this validation procedure is to detect the potential accuracy and precision achievable. Therefore, a comparison between space resection (SR) and bundle adjustment solution (BA) was carried out.

BA is a well-known numeric method to compute a multi-images position using tie-points and ground control points; this technique assures high reliability and integrity of the solutions.

Validation procedure has been conducted following these four steps:

1. Designing and building of landing pattern, using two type of circular target: ring-coded circular target and colored circular target.
2. Photogrammetric survey of landing pattern following the classical procedure employed in close-range photogrammetry.
3. Camera positions and orientations by BA using all circular target and by SRS using only colored circular target.
4. Comparison of position and orientation for each camera, between BA and SRS results.

5.4.1 *Design of Landing Pattern*

Landing pattern is composed by five circular targets attached on an aluminum plate.

In order to obtain high reliability and accuracy during the validation procedure, about 30 circular ring-coded targets were attached both on the plane face and on the three raised points (Fig. 5.9). Although the SR is not affected, in terms of solution stability, by the planar arrangement of the targets, the BA provides more reliable and precise results if the targets are located in the space rather than on the plane. A cross at the center of the circular target allows to measure the distance from a point to another with a simple caliper. Indeed, the BA in free-network adjustment (none constraints on the solution) determines positions and orientations of the cameras into an arbitrary reference system and up to scale factor. The measured distance is then applied to the solution in order to lock the photogrammetric model scale.

5.4.2 *Photogrammetric Survey*

Photogrammetry is a technique for obtaining 3D models of an object, starting from a dataset of images. Operational stages for a photogrammetry survey can be summarized in three steps:

Fig. 5.9 Coded circular targets, useful for the camera calibration and landing path referencing

1. Image acquisition, performed in the field, is based on taking a photo with a calibrated camera.
2. Object measurements, also performed in the field, are necessary to scaling the photogrammetric model.
3. Image measurement, performed in the laboratory, during this step the operator has to recognize homologues points on at least two images.

Image acquisition can be performed with a consumer camera, although it should be early calibrated. The acquisition procedure was performed following the classical guidelines for a photogrammetric close-range survey: the 3D network of images should be optimized according to the precision, reliability, and accuracy of the measurements. The 3D network realized for this work is depicted in Fig. 5.10.

The crosses, located at the center of circular targets, allow obtaining easily precise measurements of distances between the targets. A caliper was used to measure the distance between two targets (rather, between the crosses associated), in order to give a correct scale to the photogrammetric model.

The image measurements are practically fully automated, and they can be carried out employing object coded targets or natural texture of the object. Coded targets offer a subpixel image point measurement, which affords to obtain high accuracy in the course of the next stages. The precision of image coordinates point can be up to 1/50 of a pixel, yielding typical measurement precision on the object in the range of 1:100.000–1:200.000 (Brown 1971). The first one (the lower) corresponds to a precision of 0.1 mm for an object of 10 m.

In this chapter, the measurement of image coordinates targets was carried out employing two different methods:

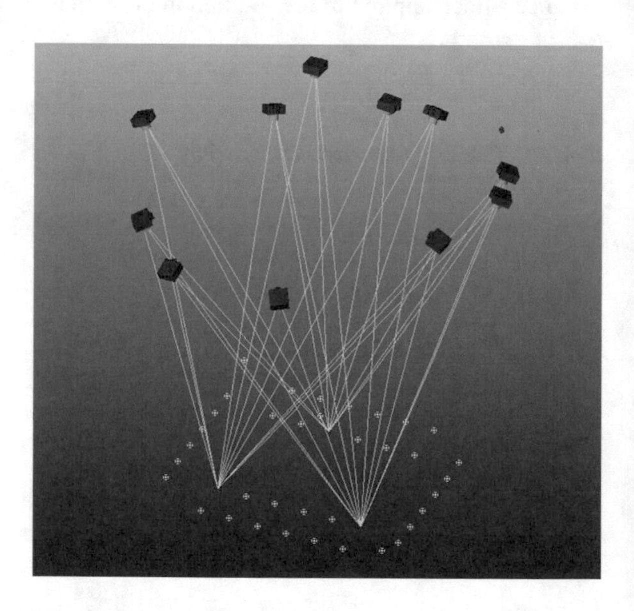

Fig. 5.10 3D camera network of the photogrammetric survey

- LSM: Least-square matching is a powerful technique for all kinds of data matching problem. Here, its application to image matching was used through the software Photomodeler Scanner [15]. Due to the perspective angle, the circular targets present on the scene could appear as an ellipse on the image. The LSM is able to detect the correct center because it applies an affine transformation to recognize the shape correctly. Such approach achieves subpixel precision, and it is widely employed in industrial photogrammetry. Unfortunately, the LSM is not available in the OpenCV library.
- CHT: Circle Hough Transform is a feature extraction technique for detecting circles on images. This method is based on the application of the Hough Transform (Luhman et al. 2011) to edge map. The latter is obtained applying the Canny edge detector (Canny 1986) with automatic thresholding on a single image. This approach is suitable for automatic recognition in soft real time, although the center of the circle is not always accurate due to the perspective angle of view, as previously described. Furthermore, it is developed in the OpenCV library.

5.4.3 Camera Orientation

The orientation is the fundamental step of the photogrammetric procedure; in this chapter, two types of orientations were performed: bundle adjustment (BA) and space resection (SRS). The former is generally performed in the laboratory, and it takes as input a large image dataset and object measurements; the output is the external orientation parameters of the entire dataset of images. In photogrammetry, the term "orientation parameters" expresses both the position and aspect of a camera. The bundle adjustment was performed using all circular target available on each image. This computation was carried out in post-processing with the commercial software Photomodeler.

On the other hand, each image orientation parameter was determined with the SRS, employing only the colored circular target and object measurements. The algorithm of SRS was developed in Matlab environment, and it is a hybrid procedure between the direct solution and the iterative one. In the first instance, the algorithm searches the three center targets visible on the generic image, so that they forming a triangle with greatest possible area. Such points are used to determine the external orientation parameters solving the problem with a non-iterative solution. This method solves the position of the camera by Ferrari's solution of quartic equations, and at a later stage, the orientation based on algebraic techniques. The possible solutions can be up to four, but usually, the real solutions are two. In order to solve the ambiguity, the iterative method is performed using as estimation of initial parameters the solution obtained with the direct method. Both the convergence factor and a statistic analysis on the re-projection residual provide a robust index to solve the ambiguity.

The non-iterative method allows speeding up the convergence of the iterative approach, obtaining a stable solution with high attitude angles also.

5.4.4 *Comparison*

The comparison is the last and fundamental step to compute the precision and accuracy achievable.

The results obtained with the bundle adjustment procedure are more robust than those of the space resection, because the redundancy of the bundle adjustment is too high, and the network is well designed. In order to evaluate the results accuracy, the BA solution is taken as reference.

5.5 Results

As reported in the previous section, in order to estimate the achievable precision and accuracy, a comparison between the BA solution and SRS one was carried out. The differences are reported in terms of mean, standard deviation, and RMSEs (root mean square errors) for each image acquired, considering two different methods of target detection: LSM and CHT.

Of course, first, a deep inspection of the photogrammetric survey is indispensable.

The survey was carried out using 14 images taken at a distance of 1.5 m and using about 37-coded circular target. The obtained network is showed in Fig. 5.10, which assures an intersection angle of 67° in average among all rays; furthermore, each image is covered by at least 28 points. Targets coordinates are estimated with high precision: indeed, the overall RMS vector length in 3D is 0.082 mm, while each camera precision values are reported in Table 5.2.

The solution provided by the bundle adjustment procedure is quite precise and robust; therefore, a comparison with SRS solution was performed.

Two different approaches were carried out; in both cases, it was considered only five circular targets. A first approach was executed using the LSM to determine the image coordinates of the targets, while the second approach was performed using the CHT method for target detection. The first one provides goods results almost in every image especially when the image is much tilted, because it is able to determine the correct target center even if the circular target appears on the image as an ellipse.

The comparison among the BA solution and both SRS solution is described in Tables 5.3, 5.4, and Fig. 5.11, where the absolute differences in 3D space of cameras position and the angle differences are reported.

Table 5.2. Values of precision about the camera position and attitude, done a classic BA procedure, for each image

Name	Precision in mm			Precision in degrees		
	X	Y	Z	ω	φ	κ
C1	0.35	0.36	0.26	0.0239	0.0226	0.0092
C2	0.34	0.33	0.20	0.0236	0.0238	0.0086
C3	0.32	0.28	0.22	0.0212	0.0197	0.0099
C4	0.32	0.35	0.21	0.0250	0.0214	0.0090
C5	0.27	0.28	0.19	0.0219	0.0197	0.0087
C6	0.26	0.32	0.24	0.0235	0.0203	0.0098
C7	0.24	0.32	0.27	0.0230	0.0186	0.0110
C8	0.22	0.25	0.26	0.0211	0.0165	0.0120
C9	0.28	0.24	0.26	0.0190	0.0170	0.0110
C10	0.34	0.38	0.26	0.0250	0.0222	0.0105
C11	0.30	0.34	0.23	0.0231	0.0223	0.0086
C12	0.32	0.32	0.20	0.0231	0.0232	0.0090
C13	1.61	1.60	0.32	0.0584	0.0584	0.0092
C14	1.21	1.19	0.21	0.0504	0.0510	0.0090

Table 5.3. LSM solution

Name	LSM detection			
	ω (°)	φ (°)	κ (°)	3D (mm)
C1	0.002	−0.014	−0.034	0.24
C2	−0.046	0.116	0.023	1.96
C3	0.080	0.052	0.029	1.70
C4	0.020	−0.034	−0.009	0.75
C5	0.009	0.110	0.014	1.57
C6	0.124	−0.124	0.064	2.71
C7	0.046	−0.037	−0.032	1.05
C8	−0.059	−0.102	0.008	2.08
C9	−0.321	−0.083	−0.004	5.16
C10	−0.027	0.077	0.009	1.25
C11	0.026	0.016	0.045	0.55
C12	−0.157	−0.134	−0.057	10.69
C13	−0.008	0.046	0.082	1.31
C14	−0.032	0.084	0.061	2.11

The histogram depicted in Fig. 5.11 shows the differences among the 3D position of perspective center computed by BA and SRS. Specifically, the differences using the HGH approach are reported in green, while LSM in blue.

Table 5.4. CHT solution

Name	CHT detection			
	ω (°)	φ (°)	κ (°)	3D (mm)
C1	1.588	−2.615	−0.946	53.84
C2	0.044	−1.190	−0.732	15.67
C3	−3.075	−0.331	0.365	43.25
C4	0.404	−1.298	0.010	22.43
C5	−1.287	0.536	−0.086	18.77
C6	0.365	−1.063	0.110	16.95
C7	0.021	0.137	0.188	4.92
C8	0.524	0.428	0.269	12.00
C9	0.591	−1.732	−0.189	32.98
C10	−0.885	1.275	−0.429	26.75
C11	−0.171	−0.232	0.097	7.44
C12	0.048	0.009	0.045	1.30
C13	−1.643	−0.342	0.593	38.24
C14	0.269	−0.461	−0.940	35.66

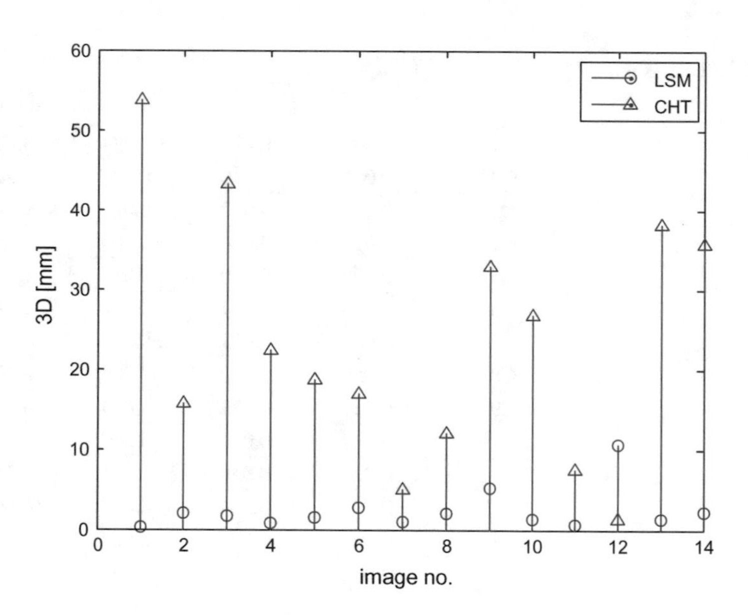

Fig. 5.11 Error differences between LSM and CHT solutions

5.6 Conclusion

A novel system for UAS landing, based on vision navigation, was presented in this paragraph. The aim of this work is to validate this system using real data.

The inspection has been highlighted, as the methodology proposed is able to achieve high precision (under the centimeter) during the landing approach phases (for about 2 m distant). Precisions cannot be achieved using a low-cost position device like a GNSS (global navigation satellite system) receiver.

This study has surveyed the LSM approach that provides more accurate results than HGH.

Chapter 6
UAS Endurance Enhancement

6.1 Introduction

Most civilian uses of UASs require the air vehicle to fly at speeds lower than 50 kts (70 km/h) and at low heights, and many applications need the ability of the aircraft to hover (e.g., power line inspection, subsurface geology, mineral resource analysis, or incident control by police and fire services) (Chap. 1). Moreover, in some cases, this type of platform needs to execute extended missions with significant flight duration time. Increasing endurance generally comes at a cost in terms of fuel consumption and airframe complexity, resulting in reduced efficiency of payload and/or range for size, mass, and financial cost.

This chapter focuses on the flight time of the vehicle and proposes an alternative approach to improve the endurance of a non-expensive, commercial quadrotor, by applying a balloon to reduce weight and power consumption needed for flight. The small quadrotor Conrad Quadcopter 450 ARF (Appendix A) has been equipped with a payload composed of a global positioning system (GPS) receiver module, an inertial measurement unit (IMU), a sonar altimeter, a small camera module, and main microcontroller hardware (Arduino, Raspberry). Typical endurance is less than 2 h (without any payload) (Wang et al. 2013), whereas in our case (motor propeller plus the payload), the flight time has found to be less than 1 h.

Extended flight duration has been mainly addressed by means of optimization of the electric propulsion system, use of hybrid-electric systems, investigation on innovative power sources, use of solar-powered HALE, or use of a variable pitch propeller. This work proposes an alternative approach to improve the endurance of a non-expensive (micro-UAV), commercial quadrotor, by applying a balloon to reduce weight and power consumption.

Some researches are available in the literature in the field of low-altitude, low-speed UAVs equipped with balloons, except for some work related to subsurface geology detection with a hybrid UAV equipped with an airbag. This approach aims for a multi-mission platform with dedicated payload, suitable for a broad range of

U. Papa, *Embedded Platforms for UAS Landing Path and Obstacle Detection*,
Studies in Systems, Decision and Control 136,
https://doi.org/10.1007/978-3-319-73174-2_6

applications in the fields of simultaneous localization and mapping (SLAM), exploration, search and rescue, remote sensing, and environmental monitoring.

The mini quadrotor used for our investigations is the Conrad Quadcopter 450 ARF (Appendix A).

The quadrotor is equipped with a payload composed of a global positioning system (GPS) receiver module, an inertial measurement unit (IMU), a sonar altimeter, a small camera module, and main microcontroller hardware (Arduino, Raspberry). Such a payload was selected considering a multi-mission data gathering platform (attitude measurement, GPS data collection, remote sensing, vision-based navigation, etc.) (Papa et al. 2014, 2015; Timmins 2011). Typical endurance of the chosen quadrotor is less than 2 h (without payload), whereas in our case (motor propeller plus payload), the flight time has found to be less than 1 h (Papa et al. 2014). Balloons are not easily maneuverable, but a hybrid solution (quadcopter merging) could effectively exploit the advantages of a quadrotor vehicle (flexibility, well-designed structure, and security) and the strength points of a balloon (low noise, low energy consumption, and buoyancy providing most of the flight lift), increasing endurance by reducing power consumption (Carson et al. 1971).

The preliminary design involves the determination of weights, gross static lifting capability of the balloon (i.e., the portion of the balloon's total lift attributable to its buoyancy), lifting gas properties, atmospheric conditions in the flight range, and influence of the balloon size on the lift. Although the dynamics and aerodynamics of the HUAS operations must be considered in a complete design of the balloon, we will consider only the balloon static performance, following the approach described in Barton (2008).

The following nomenclature is adopted for next sections:

L = static lift;
W = weight;
ρ_g, ρ_{air} = gas density, air density;
V_g = gas volume;
g = gravitational acceleration (9.8066 ms^{-2});
p = pressure;
p_g, p_{air} = gas pressure, air pressure;
R = perfect gas constant (8.314 Pa m^3 mol^{-1} K^{-1}); and
T_g, T_{air} = gas temperature, air temperature.

6.2 HUAS Conceptual Design

The most interesting improvement created by the HUAS is the aerostatic lift provided by the balloon, which, combined with the fan lift of the propulsion system, allows the vehicle to achieve easy takeoff, climbing, hovering, and landing with reduced power consumption. The main issues which inspired us to propose a hybrid solution are as follows:

- Effective control of the HUAS flight path by static lift (provided by the balloon) and rotor power. This allows hovering, flying, climbing, and landing at any height, improving the flexibility of the mission and enlarging the range of applicability.
- Low operating speed, low-altitude missions, and operation in discontinues trajectories. The HUAS could be effectively used in high-resolution spatiotemporal sampling applications and in monitoring known environments.
- Minimization of complexity of the fuselage structure, drive mechanisms, and engine systems. This implies simpler manufacturing process and shorter production cycle.
- Low noise, low vibration, and low turbulence generation. The HUAS does not disturb the environment that is being monitored or measured, and reduces noise and potential hardware malfunction due to vibration.
- Reduced cost of energy and power systems, rapid prototyping. The HUAS is an advanced low-cost system easily viable for potential commercial operators.

A stand-alone balloon is not easily maneuverable mainly for its big inertia, but in static conditions, it is very reliable. It is possible to manage and control the balloon by means of propeller speed changes. Creating hybrid lift (static lift and fan lift) to achieve takeoff, hovering, and landing with reduced energy consumption would improve flight duration. The overall system is able to correct any change of flight attitude due to voluntary actions or instability induced by the balloon.

Figure 6.1 shows the structure of the HUAS, with a basic balloon, support lines, safety lines, and a synthetic model of the quadcopter. The balloon diameter has been chosen to be 1.5 m. Thrust propellers improve the maneuverability of the HUAS. A simple web frame is chosen to wrap the balloon and link it to quadrotor; it is a nonrigid solution, able to keep the two systems linked but independent, each

Fig. 6.1 HUAS design concept

Fig. 6.2 LiPo battery packs used in research tests (4000 and 1800 mAh)

Table 6.1 Endurance for the LiPo batteries previously considered

Capacity of LiPo battery (mAh)	Takeoff and landing (min)	Hovering (min)
4000	<5	≈40
1800	<5	≈15

with its own function: the balloon for static gross lift (mainly upward) and the quadrotor for flight control. The buoyancy of the balloon provides most of the flight lift, reducing power consumption and increasing endurance. 3D CAD software was used to make a preliminary structure for successive balloon size validation.

The installed battery packs are lightweight LiPo (lithium polymer), with capacities of 1800 mAh and 4000 mAh @11.1 V, respectively (Fig. 6.2).

Endurance in minutes, without the balloon, for a normal mission (takeoff, hovering, and landing) is shown in Table 6.1.

The next paragraph focuses on HUAS weight estimation and balloon sizing. Each configuration was checked through a static analysis in CAD software, eventually choosing an optimal configuration achieving a good weight/lift compromise.

6.3 Weights Estimation and Balloon Sizing

The analytical techniques described in (Barton 2008) were applied, considering three subsystems of the HUAS:

- inflation gas;
- balloon structure;
- quadrotor system.

The weights and the static performance of each part were determined separately and successively summed. In this conceptual design phase, inertia and dynamic properties were not considered, postponing their evaluation in future work.

6.3.1 *Takeoff Weight Estimation*

For a correct evaluation of the gross static lift provided by the balloon, it was necessary to estimate the gross takeoff weight (W_{TO}) of the HUAS, a quantity which depends on all the components of the HUAS (the quadrotor structure, payloads (sensors on board), battery pack, support lines, safety lines, tether lines, and the balloon itself).

The net force acting on the balloon tether line equals the gross static lift of the balloon minus the constant tare of the bag. Equation (6.1) gives the gross static lift of the balloon (Barton 2008):

$$L = (\rho_{air} - \rho_g)V_g \tag{6.1}$$

where V_g is the volume occupied by gas when the balloon is fully inflated. In ideal conditions, the gross lift is equal to W_{TO}, given by

$$W_{TO} = W_{eq} + W_B + W_{batt} + W_p \tag{6.2}$$

where weights are referred to empty quadcopter W_{eq}, balloon (lines, gas, and bag included) W_B, battery pack W_{batt}, and payload (sensors on board) W_p. Obviously, W_B depends on the balloon size and is the unknown parameter, whereas the other contributions to W_{TO} are constant and known. Using a weight coefficient $k_x = W_x/W_{TO}$ [with x representing any of the subscripts in the right side of Eq. (6.2)], Equation (6.2) can be rewritten as

$$k_{eq} + \frac{W_B}{W_{TO}} + k_{batt} + k_p = 1 \tag{6.3}$$

And derive W_{TO} as follows:

$$W_{TO} = -W_B/(k_{eq} + k_{batt} + k_p) \tag{6.4}$$

The UAS system considered in this work has a total weight (W_{eq} + W_{batt} + W_p) of about 3 kg: this was, according to Archimedes' law, the buoyancy requirement.

6.3.2 *Ballon Static Performance and Sizing*

A classical airship has various configurations: flabby balloon with ballonet (air chamber), partially inflated, and fully inflated (Fig. 6.3). It can also be composed of two parts: a hull and a tail fin assembly. In this work, a fully inflated configuration (logging balloon, without tail fin) was chosen.

For safe flight conditions, the gas volume in fully inflated configuration must always be less than or equal to the total volume. Correct sizing must also take into account temperature and pressure influences on the lifting gas, which obviously must have density lower than air density. Among several available types of lifting gasses (hot air, hydrogen, helium, ammonia, etc.), we have chosen helium (He), due to its availability and readiness to use without many control systems (valve, pipes, etc.). Table 6.2 compares lifting forces of He and hydrogen.

From Eq. (6.1), we consider a lifting force for 1 m^3 of gas at sea level and 0 °C. Lift decreases proportionally with the altitude (temperature and pressure). The evaluation of the lifting force is only referred to 1 m^3 of lifting gas and does not include the bag and lines of the balloon.

The balloon was modeled as a sphere, which contains lifting gas sufficient to equalize, more or less, W_{TO}. The main mission parameters to be defined are as follows: altitude (pressure altitude—*PA*, temperature *T*), gas density, and air density. Typical *PA* values in the range 6–20 m (20–65 ft) and air temperature of 288 K (15 °C), reasonably constant in the selected altitude range, were considered. The density–pressure–temperature nomogram (Fig. 6.4) can be used to determine the inflation requirements of the balloon.

The blue dashed line in Fig. 6.4 shows the initial inflation in the selected operational conditions and determines a balloon gas density of 0.168 kg/m^3. The design volume V_D of the spherical balloon and the weight of the inflation gas, W_{He}, were computed through use of the perfect gas law, giving

$$V_D = \frac{W_{TO}}{\left(1 - \frac{p_g}{p_{air}} \frac{R_{air}}{R_g}\right) \frac{g p_{air}}{R_{air} T^*}} \tag{6.5}$$

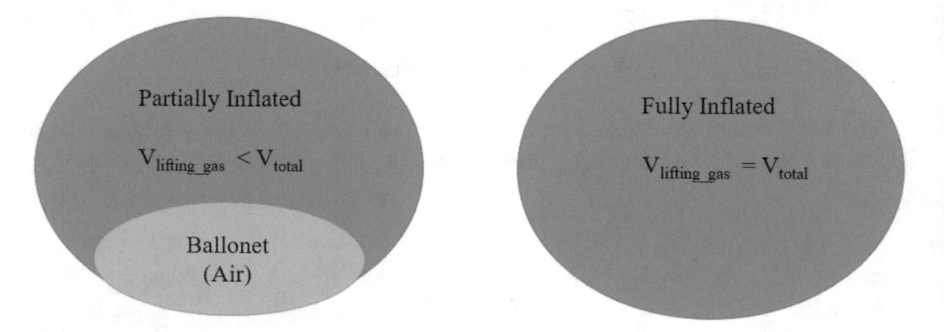

Fig. 6.3 Balloons configuration

Table 6.2 Comparison of lifting gas properties

Lifting gas	Density at sea level and 0 °C (kg/m^3)	Lifting force of 1 m^3 of gas (N)
Helium (He)	0.178	11.8
Hydrogen (H)	0.090	10.9
Air	1.292	##

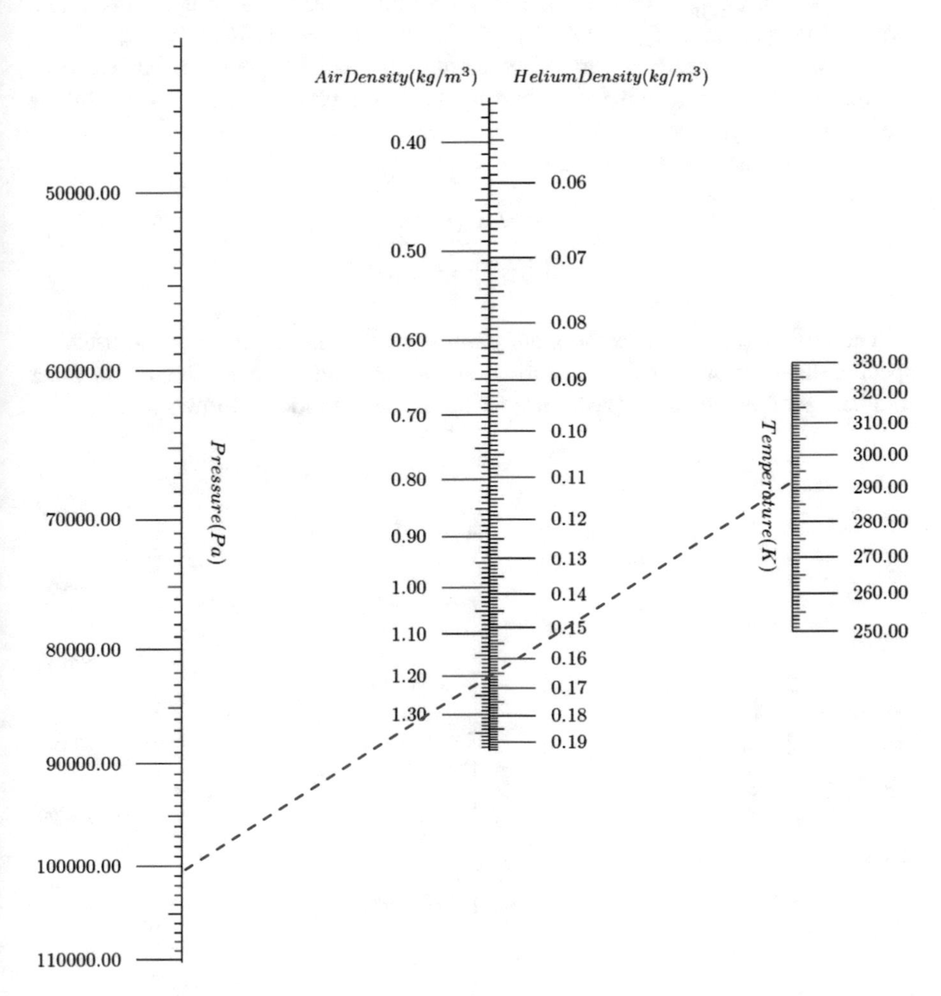

Fig. 6.4 Density–pressure–temperature nomogram and initial inflation determination (dashed line)

$$W_{He} = \rho_g V_D \tag{6.6}$$

where $T^* = T_g/T_{air}$ and V_D is referred to a fully inflated balloon.

Equation (6.7) gives the design volume (cubic feet is the preferred unit in the compressed-gas industry) required in standard conditions, V_{STD}:

$$V_{\mathrm{STD}} = \frac{\rho_g}{\rho_{\mathrm{STD}}} \frac{W_{\mathrm{TO}}}{\left(1 - \frac{p_g}{p_{\mathrm{air}}} \frac{R_{\mathrm{air}}}{R_g}\right) \frac{g p_{\mathrm{air}}}{R_e T^*}} \tag{6.7}$$

It is now possible to calculate the expected gross static lift (L_{exp}), given by Eq. (6.1), considering the nomogram for initial inflation, shown in Fig. 6.5. From the blue dashed line, L_{exp} is found to be 6.78 kg per kg of He, considering the maximum expected temperatures: $T_{\mathrm{air}} = 299$ K (≈27 °C) and $T_{\mathrm{He}} = 322$ K (≈49 °C).

Generally, L depends on atmospheric pressure and temperature, but we considered a constant pressure value, due to the low altitude and the small altitude range (0–19 m) chosen.

The gross static lift L is given by

$$L = \frac{L_{\mathrm{exp}} \rho_g W_{\mathrm{TO}}}{\left(1 - \frac{p_g}{p_{\mathrm{air}}} \frac{R_{\mathrm{air}}}{R_g}\right) \frac{\rho_{\mathrm{STD}} g p_{\mathrm{air}}}{R_{\mathrm{air}} T^*}} \tag{6.8}$$

The design volume V_{D} is the main parameter determining gross static lift for a specific diameter of the balloon. Various diameters have been considered, in order to attain good efficiency ($L(V_{\mathrm{D}})/Drag \approx 1$), at low altitude and low speed.

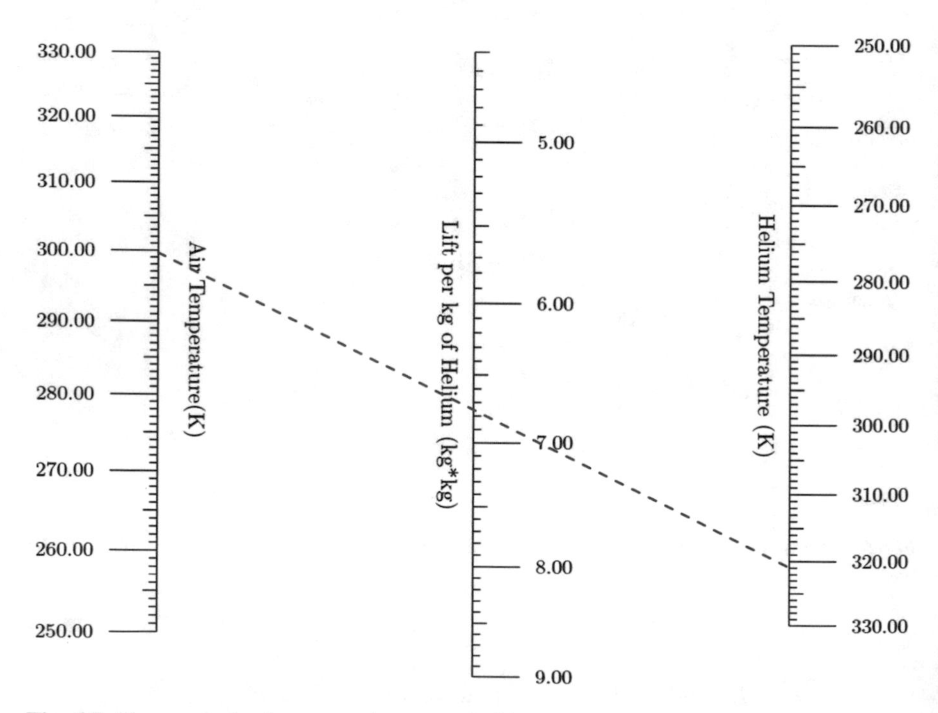

Fig. 6.5 Nomogram for the expected gross static lift ($T_{\mathrm{air}} = 299$ K, $T_{\mathrm{He}} = 322$ K)

Recalling that W_{TO} is equal to 3 kg, the balloon weight W_B, given by

$$W_B = W_{He} + W_{bag} + W_{lines} \tag{6.9}$$

has to be added. W_{He} is given by Eq. (6.6), and the remaining weights are a function of VD and the material used for the bag and the lines.

The balloon thickness is obviously dependent on the helium pressure, which in turn depends on temperature, volume of the bag (VD), and number of moles of He. Using the ideal gas law and assuming operational conditions at T = 15 °C (288.15 K) and p = 1031.25 hPa, the helium pressure, which must be sustained by the structure (bag), is given by

$$p_{He} = \frac{nRT}{V_D} \tag{6.10}$$

Young–Laplace law (Batchelor 2000) can be used to derive the pressure difference $\Delta p = p_{He} - p_{air}$, across the interface between air and helium, assuming a spherical balloon of radius r:

$$\Delta p = \frac{2\tau(r)}{r} \tag{6.11}$$

where τ is the surface tension (obviously, Δp should be less than the maximum tension of the balloon, to avoid bursting). The material chosen for the balloon was PVC (polyvinyl chloride), whose mechanical characteristics are summarized in Table 6.3.

It turned out that the tension forces τ are small and the balloon pressure is very close to the atmospheric pressure. Therefore, thickness in the range 0.10–0.30 mm is sufficient to guarantee the necessary lift, keeping the balloon away from bursting conditions.

6.4 Preliminary Results

The test field and the atmospheric conditions are very important in order to evaluate the lifting properties of the inflation gas used for our HUAS. Experimental results were acquired considering three cases (modeling three different environmental conditions):

Table 6.3 PVC mechanical characteristics

Property	Units	Method	Value
Specific weight	g/cm^3	ISO 1183	1.42
Yielding tension	MPa	DIN EN ISO 527	58
Elastic modulus	MPa	DIN EN ISO 527	3000
Hardness SHORE D	–	ISO 868	82

- Standard operating conditions (International Standard Atmosphere, ISA, 0 km mean sea level (MSL), temperature 15.0 °C (288 K), density 1.225 kg/m^3, and pressure 1013.25 hPa);
- Complex scenario (ISA+25 and ISA−25).

In this case, only the temperature variation of air and of the inflation gas is considered. Relative humidity is not taken into account due to its negligible influence on the lifting properties. Airspeed has been set equal to 0. No variation on the payload mass was taken into account, since there is no fuel consumption. Table 6.4 summarizes the expected operating conditions and the helium expected gross static lift in the standard operating conditions.

Helium density and the expected gross static lift are quite similar (see Figs. 6.4 and 6.5) and were considered constant. On average, we have a gross static lift [from Eq. (6.8)] of 2.84 kg. In this case, one kg of He is sufficient to lift the UAS and sensors ($W_{eq} + W_{batt} + W_p$).

Using a spherical configuration, the total gross static lift becomes a function of the balloon diameter, as shown in Fig. 6.6 [see also Eq. (6.8)].

Figure 6.7 shows the dependence of the gas weight (to be added to obtain W_{TO}) on the balloon diameter.

In order to analyze the effects of temperature on the static lift, we performed several tests in different conditions, namely, ISA−25 and ISA+25, i.e., −10 °C and +40 °C, as operational boundaries in which the HUAS is expected to work.

By using the nomograms of Figs. 6.4 and 6.5 and Eq. (6.8), an estimate of the expected helium gross static lift is shown in Table 6.5 and Figs. 6.8 and 6.9.

Neglegible differences were found with respect to the standard condition, concluding that the balloon can provide the required lift even in complex scenarios, since its inflation requirement is not dependent on temperature in the range from −10 to +40 °C.

Total balloon weight W_B is estimated considering a PVC bag with thickness in the range 0.18–0.28 mm. The weight of the link lines is reasonably constant within the considered thickness range.

A diameter of 2.7 m provides full W_{TO} equivalence (bag and lines included, see Fig. 6.7), but in terms of control and maneuvering, it is a poor choice, since such a value involves high drag force, depending on body sectional area, acting opposite to the relative motion. This, in turn, involves weak maneuverability and control. Since

Table 6.4 Lifting properties of He in standard (ISA) operating conditions

Pressure altitude (m)	Air or helium temperature (°C)	Helium expected gross static lift (kg per kg of He)	Helium density (kg/m^3)	Air density (kg/m^3)
7	17.1045	6.78	0.1681	1.2241
10	17.0850	6.78	0.1681	1.2238
13	17.0655	6.78	0.1681	1.2234
16	17.0460	6.78	0.1682	1.2231
19	17.0265	6.78	0.1682	1.2227

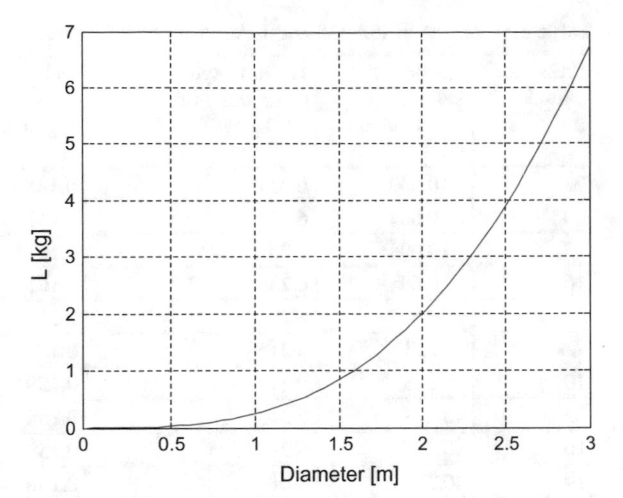

Fig. 6.6 Gross static lift versus balloon diameter

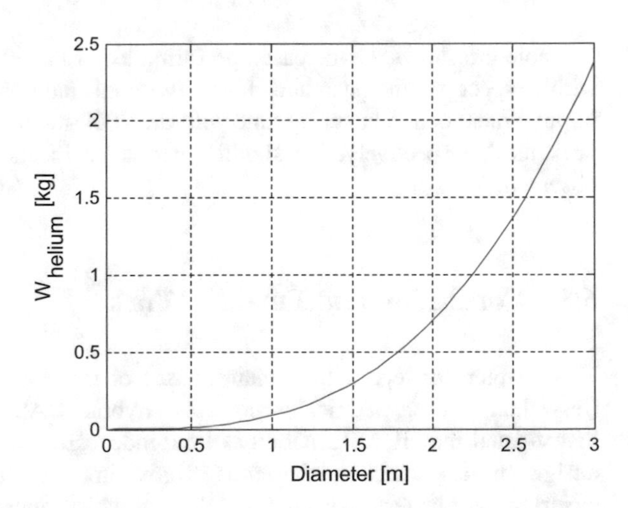

Fig. 6.7 Gas weight versus balloon diameter

our goal is to reduce power consumption of the rotors to increase endurance, it is not necessary to maximize the total gross lift (i.e., $L = W_{TO}$). In this preliminary design (Fig. 6.10 shows the relationships among design parameters), a solution with gross static lift less than W_{TO} has been chosen, resulting in low diameter and low drag force, with acceptable maneuverability and control.

Under these hypotheses, considering the initial inflation parameters, it is be possible to estimate gross static lift values for a specific balloon diameter. The generated gross lift is 67% of W_{TO}, for a 2-m balloon diameter: in this condition, the HUAS is able to come back to the ground by its own gravity. During landing and hovering, the lift provided by the balloon and quadcopter can help to get the desired hovering height; so, the total lift is less than the HUAS weight (sensors included). If the propeller speed decreases, the HUAS can land due to its own weight.

Table 6.5 Results of ISA−25 and ISA+25 operating conditions

Pressure altitude (m)	Air or helium temperature (°C)	Helium expected gross static lift (kg per kg He)	Helium density (kg/m^3)	Air density (kg/m^3)	Gross static lift (Eq. 6.8) (kg)
7	−10.609	6.23	0.161	1.17	2.61
10	−10.609	6.23	0.161	1.17	2.61
13	−10.608	6.23	0.161	1.17	2.61
16	−10.605	6.23	0.161	1.17	2.61
19	−10.605	6.23	0.161	1.17	2.61
7	40.253	6.21	0.158	1.15	2.61
10	40.248	6.21	0.158	1.15	2.61
13	40.233	6.21	0.158	1.15	2.61
16	40.230	6.21	0.158	1.15	2.61
19	40.226	6.21	0.158	1.15	2.61

Table 6.6 shows the increased performance of the HUAS in the hovering phase, with respect to the standard UAS (without balloon), showing 50 and 40% improvement on the hovering time with the 4000-mAh and 1800-mAh LiPo battery pack used, respectively. No significant change in takeoff and landing time was observed.

6.5 Conclusion and Further Work

This chapter has reported the main phases of the conceptual design of a low-cost (less than 2 k€), electrically powered hybrid UAS (quadrotor + airship). We investigated the HUAS capability of extended cruise endurance by analyzing the design drivers affecting the craft flight time. With respect to conventional electric-motor/battery powered UASs, in which increasing endurance requires heavier batteries, with a consequent weight increment, we propose a solution that increases endurance by using a balloon, resulting in a favorable endurance/weight ratio.

The installation of a balloon with a diameter of about 2 m provided a significant increase in flight time, as demonstrated in test missions developed in situ (urban traffic monitoring scenario). As shown in Table 6.6, the HUAS endurance (with 1800-mAh LiPo battery) increased by 50%. The effect of temperature changes on the static lift provided by the balloon (inflated with helium) has found to be negligible, providing an average value of 2.84-kg lift (27.8 N) per kg of He.

Further investigations will focus on stability and control problems of the HUAS, caused by wind flow (e.g., vertical/horizontal gusts), critical scenarios in takeoff and landing (e.g., sloped terrain, obstacles), and pendulum effects during left/right turning, which make trajectory tracking and attitude stabilization challenging tasks.

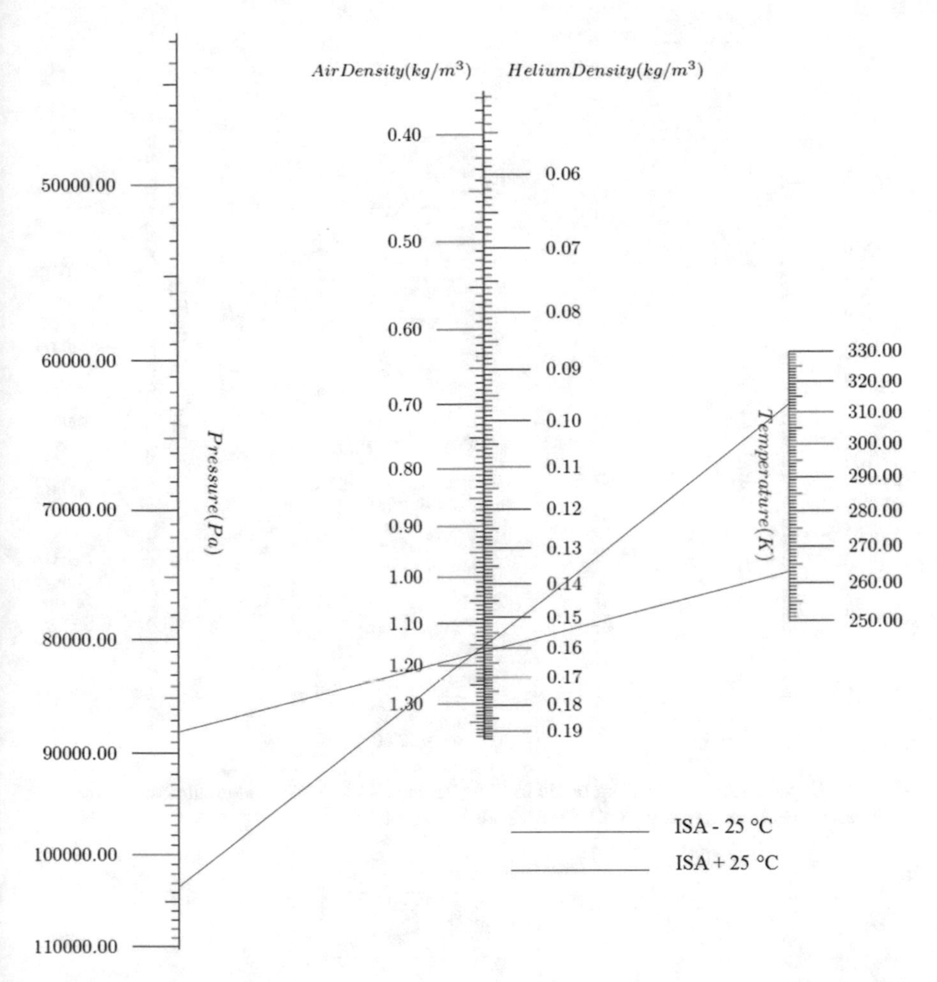

Fig. 6.8 Air and helium densities in ISA−25 and ISA+25

These effects mainly depend on nonlinearities like coupling between the quadrotor and the balloon. Nonlinear techniques are currently under study, together with a structural modification of the craft. The main theoretical aspects to be analyzed in further developments will concern optimal control techniques for the following critical issues:

- Stabilization and attitude control during the hovering phase and capability of tracking straight-line trajectories;
- Transition between flight modes and operation near ground (especially in takeoff and landing missions, where the HUAS is supposed to smoothly reach the desired hovering height, or to descend from initial height to 0 m);
- Collision avoidance.

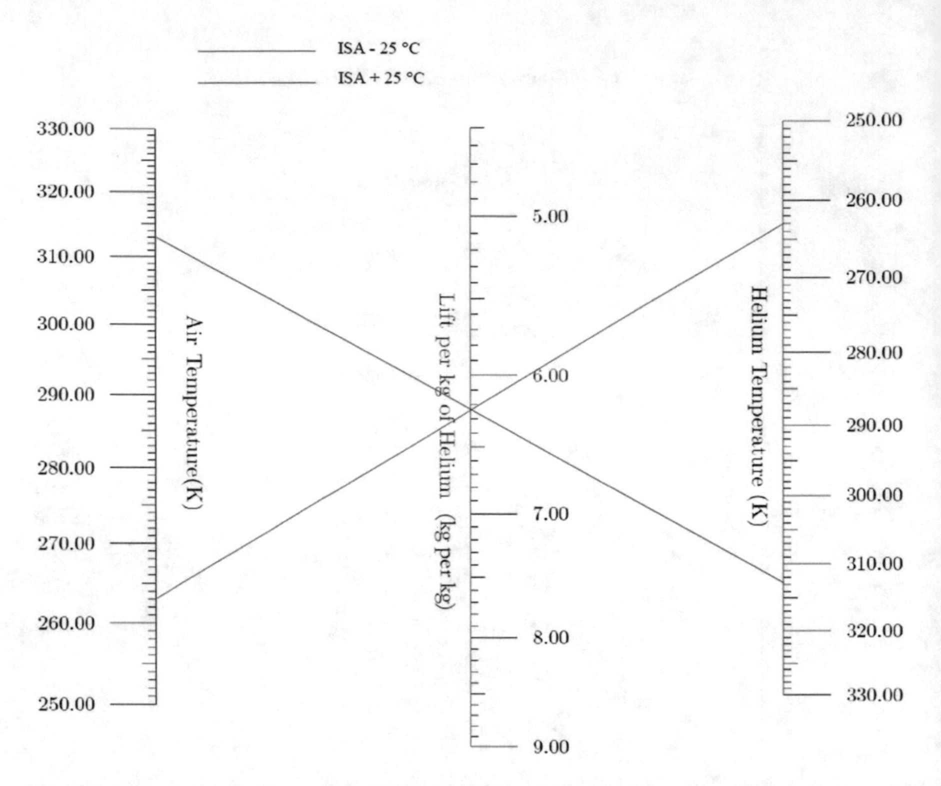

Fig. 6.9 Expected static lift per kg of He in ISA−25 and ISA+25. The value (almost the same for both cases) has been found to be about 6.22 kg per kg of He

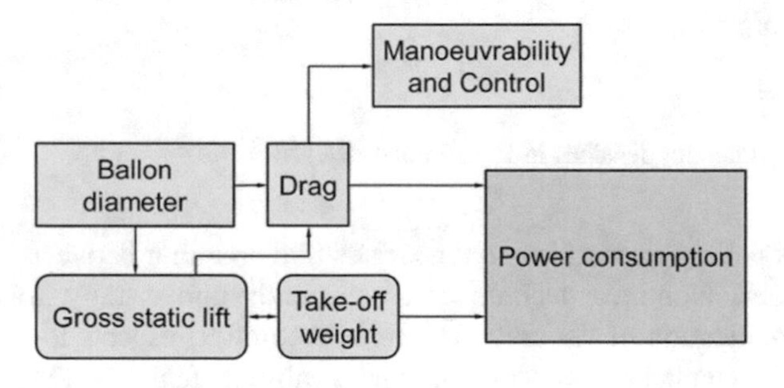

Fig. 6.10 Relationships among design parameters

Table 6.6 HUAS versus UAS endurance

	UAS		HUAS	
	Takeoff and landing (min)	Hovering (min)	Takeoff and landing (min)	Hovering (min)
4000 mAh	<5	≈40	<5	≈60
1800 mAh	<5	≈15	<5	≈25

Fig. 6.11 Latest HUAS configuration, with directional propellers

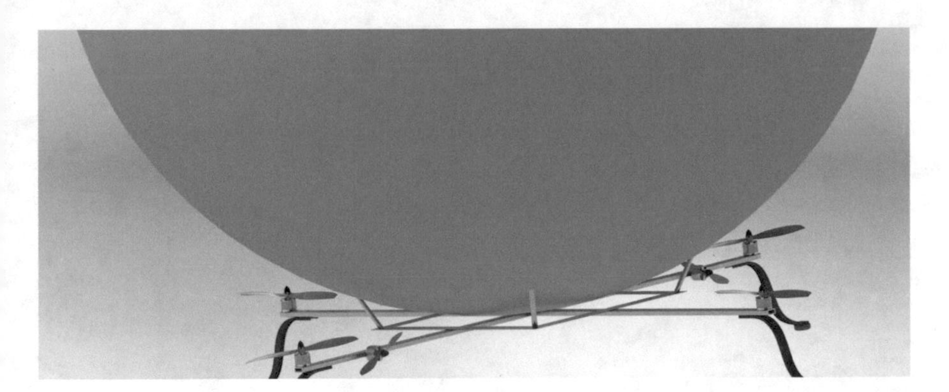

Fig. 6.12 Newest configuration to reduce nonlinearity

As a preliminary step, two additional propellers (Fig. 6.11) will be added to the existing configuration in order to provide thrust for flight direction changing. To remove the nonlinearity introduced by pendulum moments, the tether lines will be reduced, obtaining a final UAS + balloon single structure shown in Fig. 6.12. Preliminary simulations and dynamic studies confirm improved HUAS stability, allowing us to tune the mechanical design for optimal control sensitivity and disturbance rejection.

Another issue to be explored in further work is the purity of the lifting gas (it is not easy to find 100% pure He), which could impair the gas performance during the mission. The impact of non-pure helium on the lifting properties is currently under examination.

Chapter 7
Conclusions

This book wants to be a novelty contribution to the many topics covered in the UASs applications. A monitoring and control landing system was done, in order to assist the remote pilot during the landing procedure when the environmental condition is adverse. The whole system was composed of ultrasonic sensor, infrared sensor, and optical sensor; each one of them has been discussed in the chapters of this work. The UAS will be furtherly equipped with others sensors (e.g. LIDAR) in manner to enhance the landing path estimation and obstacle detection and avoidance.

According to Joint Publication 1-02, DoD Dictionary (OSD 2002), I wish to conclude this work by giving the definition of UAS or unmanned aircraft UA:

> A powered aerial vehicle that does not carry a human operator uses aerodynamic forces to provide vehicle lift, can fly autonomously or be piloted remotely, can be expendable or recoverable, and can carry a lethal or non-lethal payload. Ballistic or semi ballistic vehicles, cruise missiles, and artillery projectiles are not considered unmanned aerial vehicles.

U. Papa, *Embedded Platforms for UAS Landing Path and Obstacles Detection*,
Studies in Systems, Decision and Control 136,
https://doi.org/10.1007/978-3-319-73174-2_7

Appendix

UAVs Features

This appendix describes the main features of the UASs utilized during this work.

Conrad 450 ArF

The research conducted in Chaps. 2, 5 and 6 used as a test bed this quadrotor (CONRAD 2016), made from Reely and distributed over Conrad. The central control is based on an efficient (Atmel-Mega 128) microprocessor, which conveys the control commands to the processors for the motor electronics via a bus system.

A complex control electronics comprising of position and acceleration sensors, as well as efficient RISC microprocessors, stabilize the quadrotor during flight.

As stated in (CONRAD 2016), highlights and details were listed below:

- Flying platform for air reception;
- Up to 500 gr. additional loads possible;
- System-driven microprocessor;
- Individual programming by optional software;
- Ideally suitable for night flight (Fig. A.1 and Table A.1).

RC EYE NovaX 350

This UAS (RC Logger NovaX 2015) consolidates rigid design, modern unmanned aeronautic technology as well as easy maintainability into one outstanding versatile multirotor platform. In the professional field, quadrotors are used for the most different tasks (aero photogrammetry, search and rescue, environment monitoring, industrial component inspection, etc.). It is also possible to choose a fixed-wing UAS configuration, but the airspace is typically limited. A fixed-wing configuration needs a runway to land and take off, so rotorcraft are much flexible in these procedures.

U. Papa, *Embedded Platforms for UAS Landing Path and Obstacle Detection*,
Studies in Systems, Decision and Control 136,
https://doi.org/10.1007/978-3-319-73174-2

Fig. A.1 Several images of the 450 ArF quadrotor

Table A.1 450 ArF main dimensions (CONRAD 2006)

Dimensions (∅ × H)	450 mm × 165 mm
Main rotor ∅	260 mm
Weight	670 g
Load capacity	~500 g

The NovaX 350 is fully electric, with four propellers installed on direct current brushless motor in connection with a specially developed control, user programmable.

The NovaX 350 design is depicted in Fig. A.2.

This vehicle can be operated both indoors and outdoors during calm weather conditions (without wind). It is preferred, for be in safe, to do outdoor analysis. The built-in electronic controls can balance out small-undesired changes to the flight altitude. Aim is to control and manage this system (in particular, landing procedure, and flight mechanics main parameters), through IR+SR sensors distance acquisition.

The RC EYE NovaX 350, is classified like a small/mini UAS, according to Tables 1.1 and 1.3 (OSD 2001).

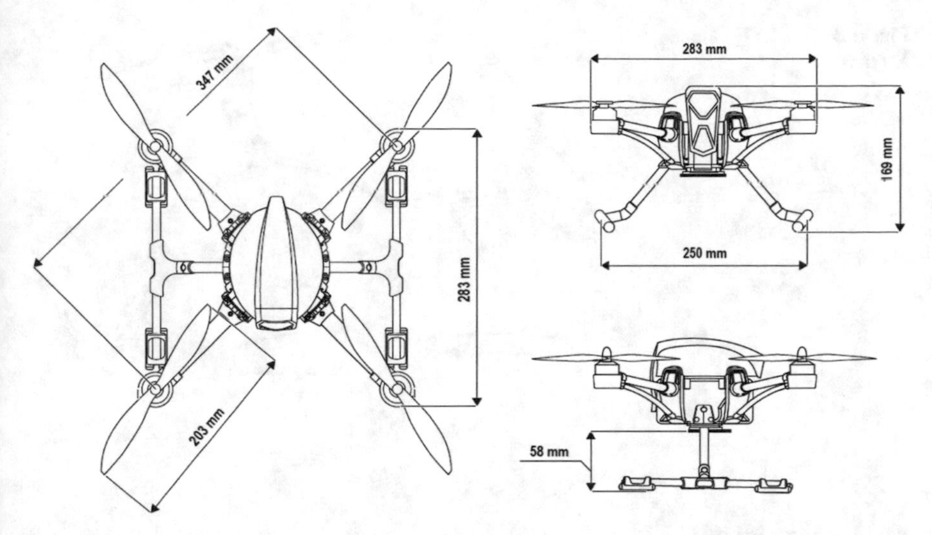

Fig. A.2 RC NovaX 350

Table A.2 Active sensors during each flight mode

Sensor	Application	Flight mode		
		Basic	Altitude	GPS
Gyroscope	Orientation	·	·	·
Accelerometer	Leveling	·	·	·
Barometer (atmosphere pressure sensor)	Altitude	–	·	·
GNSS sensor	GPS	–	–	·

The sensors available during the navigation depends on the flight mode and are described in Table A.2.

RC EYE One Xtreme

The RC Eye One Xtreme (Fig. A.3), a micro quadrotor, is a vehicle of a new series (Nonami et al. 2010; RC Logger Eye 2015). Micro quadrotors will be the future of UAS technologies.

The platform accommodates the latest 6-axis gyro stabilization technology, outstanding brushless motor driven flight control, all embedded within a robust yet stylish frame design. The sturdy lightweight construction is an ideal platform for flight applications ranging from aerial surveillance, imaging or simply unleashing acrobatic fun flight excitement. It is possible to select two flight modes: beginners

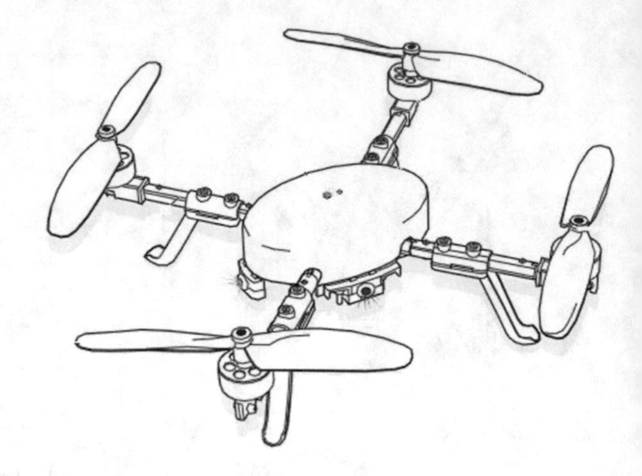

Fig. A.3 RC EYE One Xtream

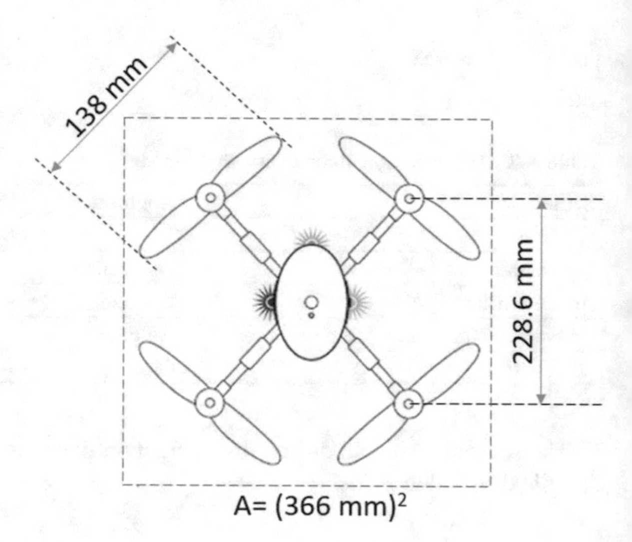

Fig. A.4 RC EYE One Xtream main dimensions

and experts flight enjoyment alike. In this work, for data session, beginner's mode is enabled (Fig. A.4).

A 2.4-GHz receiver equips quadrotor, therefore, it is possible to configure a 2.4-GHz transmitter, in order to control and manage it.

UAS use different colors (e.g., for propellers, led, legs, etc.) to distinguish the nose from the tail, this is important for the referencing during data collection. Each point acquired, need to be referenced respect a reference coordinate system. UAS data collection and measurements were evaluated in a reference system, where its origin is the centroid of the vehicle.

References

3DRAG (2016) Product description. https://store.open-electronics.org/3Drag%203D%20printer%20KIT. Accessed 13 Jan 2016

Austin R (2010) Unmanned aircraft systems—UAVs design, development and deployment. Wiley, vol 54

Arduino.cc (2016) https://www.arduino.cc/en/Main/ArduinoBoardMega2560#techspecs. Accessed 13 July 2016

Barton SA (2008) Stability analysis of an inflatable vacuum chamber. J Appl Mech 75(4)

Batchelor GK (2000) An introduction to fluid dynamics. Cambridge University Press

Brown, Duane C (1971) Close-range camera calibration. Photogramm. Eng, 37(8), 855–866

CONRAD, http://www.conrad.com/ce/en/product/208000/QUADROCOPTER-450-ARF-35-MHz. Accessed 15 Jan 2016

Climatic Chamber KK-105 (2016) Technical datasheet. http://www.kambicmetrology.com/climatic-chambers/kk105-ch, Kambic Ltd. Accessed 20 July 2016

Canny JA (1986) A computational approach to edge detection. IEEE Trans Pattern Anal Mach Intell 8(6):679–714

Carson WW, Peters PA (1971) Gross static lifting capacity of logging balloons. Pacific Northwest Forest and Range Experiment Station. Forest Service, US Department of Agriculture, vol 152

Dickerson L (2007) UAVs on the rise. Aviation week and space technology. Aerospace Source Book 2007 vol. 166, No. 3

Dean EA (1979) Atmospheric effects on the speed of sound. Technical Report of Defense Technical Information Center, Battelle Columbus Labs Durham NC

DHT11 sensor (2010) D-robotics, technical datasheet www.micropik.com/PDF/dht11.pdf, www.droboticsonline.com. Accessed 21 July 2010

Diosi A, Geoffrey T, Lindsay K (2005) Interactive SLAM using laser and advanced sonar. In: Proceedings of the 2005 IEEE international conference on robotics and automation. IEEE, pp 1103–1108

Eisenbeiss H (2004) A mini unmanned aerial vehicle (UAV): system overview and image acquisition. 2004 International Workshop on "Processing and Visualization using High-Resolution Imagery". Institute for Geodesy and Photogrammetry, ETH-Hoenggerberg, Zurich, CH

Endoh N, Tsuchiya T, Yamada Y (2003) Relation between received pulse of aerial ultrasonic sonar and temperature and relative humidity. World Congress on Ultrasonic, Paris

Farid bin Misnan M, Arshad NM, Razak NA (2012) Construction sonar sensor model of low altitude field mapping sensors for application on UAV. In: IEEE 8th colloquium on signal processing and its applications, IEEE, pp 446–450

Faugeras O (1993) Three dimensional computer vision: a geometric viewpoint. MIT Press

Fantoni I, Lozano R (2002) Non-linear control for underactuated mechanical systems. Springer Science & Business Media, London, UK

U. Papa, *Embedded Platforms for UAS Landing Path and Obstacle Detection*,
Studies in Systems, Decision and Control 136,
https://doi.org/10.1007/978-3-319-73174-2

Fraser Clive S (1997) Digital camera self-calibration. ISPRS J Photogram Remote Sens 52(4): 149–159

Gelb A (1974) Applied optimal estimation. MIT Press, Cambrige

Graham G (2006) The standard atmosphere. A mathematical model of the 1976 U.S. Standard Atmosphere

HC-SR04 (2010) Ultrasonic ranging module, technical datasheet. ITead Studio

HD2817T (2011) Manual, Delta Ohm HD281T, Transmitter, indicator, ON/OFF regulator, temperature and humidity datalogger. Rev. 1.2, 28 Nov 2011

Hough PVC (1962) Method and means for recognizing complex patterns. Patent, US3069 654, USA

Luhman T, Robson S, Kyle S, Harley I (2011) Close range photogrammetry, principles techniques and applications. Whittles Pub

Ludington B, Johnson EN, Vachtsevanos GJ (2007) Vision based navigation and target tracking for unmanned aerial vehicles. Advances in Unmanned Aerial Vehicles—State of Art and the Road to Autonomy. Springer Netherlands, pp 245–266

Mustapha B, Zayegh A, Begg RK (2012) Multiple sensors based obstacle detection system. In: 4th international conference on intelligent and advanced systems

Nonami K, Kendoul F, Suzuki S, Wang W, Nakazawa D (2010) Autonomous flying robots: unmanned aerial vehicles and micro aerial vehicles. Springer

National Instruments™ (2003) LabVIEW™ User Manual, P.N.: 320999E-01, Ed., April 2003

OSD (2001) Office of the Secretary of Defence, Unmanned Aerial Vehicles Roadmap, April 2001

OSD (2002) UAV Roadmap 2002–2027, Office of the Secretary of Defence (Acquisition, Technology, & Logistics), Air Warfare, December 2002

OSD (2005) Office of the Secretary of Defence, Unmanned Aerial Systems Roadmap 2005–2030, Technical Report

Ping Parallax (2016) Technical datasheet, PING))). https://www.parallax.com/product/28015. Parallax Inc. Accessed 21 June 2016

Picamera (2013) https://picamera.readthedocs.io/en/release-1.12/. Release 1.12. Accessed 21 July 2015

RC Logger, NovaX (2015) Operating instructions, RC EYE NovaX 350. http://www.rclogger.com/support, CEI Conrad Electronic International (HK) Limited

RC Logger, One Eye (2015) Operating instructions, RC EYE One Xtreme. http://www.rclogger.com/support, CEI Conrad Electronic International (HK) Limited

RaspberryPi camera (2016) https://www.raspberrypi.org/products/camera-module/. Accessed 16 Feb 2016

RaspberryPi (2016) Technical documentation. https://www.raspberrypi.org/documentation/. Accessed 5 July 2016

Sobers Jr. DR, Chowdhary D, Johnson EN, (2009) Indoor navigation for unmanned aerial vehicles. In: AIAA guidance, navigation, and control conference, pp 10–13

Sharp (2006) GP2Y0A02YK0F infrared sensor, Data Sheet, Sheet No.: E4-A00101EN, ©SHARP Corporation

Sharp CS, Shakernia O, Sastry SS (2001) A vision system for landing an unmanned aerial vehicle. In Robotics and Automation, 2001. Proceedings 2001 ICRA. IEEE International Conference, Vol 2, pp 1720–1727

Song K, Chen C, Chiu Huang C (2004) Design and experimental study of an ultrasonic sensor system for lateral collision avoidance at low speed. In: IEEE intelligent vehicles symposium, IEEE, pp 647–652

Timmins H (2011). Robot integration engineering a GPS module with the Arduino. Practical Arduino engineering. Springer Science+Business Media, New York, Chap. 5, pp 97–131

Tardos JD, Neira J, Newman PM, Leonard JJ (1990) Robust mapping and localization in indoor environments using sonar data. Int J Robot Res 21(6):560–569

US Army (2010) Roadmap for UAS 2010–2035, US Army UAS Center of Excellence (ATZQ-CDI-C), Forth Rucker, Alabama

Valavanis KP (2008) Advances in unmanned aerial vehicles—state of art and the road to autonomy. Springer Science & Business Media, vol 33

Valavanis KP, Kontitis M (2008) A historical perspective on unmanned aerial vehicles. Advances in unmanned aerial vehicles—state of art and the road to autonomy. Springer Science & Business Media, vol 33, pp 15–48

Valavanis KP, Vatchesevanos GJ, Anstaklis PJ (2008) Conclusion and the road ahead. Advances in Unmanned Aerial Vehicles—State of Art and the Road to Autonomy, Springer Science Business & Media, vol 33, pp 533–543

Wang Y, Le S, Fan Z, (2013) A new type of rotor+ airbag hybrid unmanned aerial vehicle. Int. J. Energy Sci (IJES) 3(3):183–187

Yi Z, Khing HY, Seng CC, Wei ZX (2000) Multi-ultrasonic sensor fusion for mobile robots. In: Proceedings of 2000 IEEE international symposium on intelligent vehicle, IEEE, pp 602–607

About the Author

Umberto Papa was born on June 20th, 1986 in Caserta, Italy; he grew up in a little town near Caserta. Driven by strong passion for aeronautics, after elementary and junior high school, he attended the Technical Aeronautic Institute achieving the High School Diploma in 2004. Umberto immediately joined the faculty of Aerospace Engineering achieving a bachelor's degree in 2008, with a thesis titled "Fatigue Evaluation on a General Aviation Aircraft". In 2012, he obtained his master's degree in Aerospace Engineering, with honors, with a thesis titled "Hardware and Software Systems Design for Reverse Engineering Techniques". He then worked as a mechanical designer for an industrial automation company. In September 2014, he became research fellow at University of Naples "Parthenope", with research topic of augmentation techniques for air navigation". On October 30, 2015, he won "Leonardo Innovation Award", in the Ph.D. student category.

In June 2017, he achieved the Ph.D. in applied science on sea, environment, and territory at the University of Naples "Parthenope"; his main research topics are focused on the flight mechanics, navigation, and guidance of small UASs.

Actually, he is an aerospace engineer at Leonardo Company—Aircraft Division.

U. Papa, *Embedded Platforms for UAS Landing Path and Obstacle Detection*,
Studies in Systems, Decision and Control 136,
https://doi.org/10.1007/978-3-319-73174-2

Author Award

This book topic was one of the winners of the Leonardo Innovation Award 2015. The ceremony was held at the Universal Exposition in Milan, EXPO 2015.

2nd Place, Ph.D. Category, Topic: Autonomous Systems

Project: PERSEO (Piattaforma Embedded per la Ricerca Sentiero di Atterraggio ed Ostacoli) (Fig. 1).

Fig. 1 Prize giving, from left with Mauro Moretti (Leonardo (ex Finmeccanica) CEO), Stefania Giannini (Minister of Education), Gianni De Gennaro (Leonardo (ex Finmeccanica) President), and Umberto Papa. October 31, 2015

U. Papa, *Embedded Platforms for UAS Landing Path and Obstacle Detection*,
Studies in Systems, Decision and Control 136,
https://doi.org/10.1007/978-3-319-73174-2

Author Publications

Most of the author's publication, below, are made from the experiences reported in this book.

Papa, U., & Del Core, G. (2014). Design and Assembling of a low-cost Mini UAV Quadcopter System. *Department of Science and Technology, University of Naples Parthenope*.

Papa, U., & Del Core, G. (2015, June). Design of sonar sensor model for safe landing of an UAV. In Metrology for Aerospace (*MetroAeroSpace*), *2015 IEEE* (pp. 346–350). IEEE.

Papa, U., Del Core, G., & Picariello, F. (2016). Atmosphere effects on sonar sensor model for UAS applications. *IEEE Aerospace and Electronic Systems Magazine*, 31 (6), 34–40.

Papa, U., Del Core, G., & Giordano, G. (2016, August). Determination of sound power levels of a small UAS during flight operations. In *INTER-NOISE and NOISE-CON Congress and Conference Proceedings* (Vol. 253, No. 8, pp. 692–702). Institute of Noise Control Engineering.

Papa, U., Ponte, S., & Del Core, G. (2017). Conceptual Design of a Small Hybrid Unmanned Aircraft System. *Journal of Advanced Transportation*, 2017.

Papa U., Ponte S. Del Core G. and Giordano G. (2017). Obstacle Detection and Ranging Sensor Integration for a Small Unmanned Aircraft System. In Metrology for AeroSpace (*MetroAeroSpace*), *2017 IEEE* International Workshop on (pp. 571–577). IEEE.

Papa U., (2017). Student research highlights: Embedded Platform for UAS Landing Path and Obstacles Detection—PERSEO. *IEEE Aerospace and Electronic Systems Magazine*, 32(7), 58–61.

Papa, U., Iannace, G., Del Core, G., & Giordano, G. (2017). Sound power level and sound pressure level characterization of a small unmanned aircraft system during flight operations. *Noise & Vibration Worldwide, 48*(5–6), 67–74.

U. Papa, *Embedded Platforms for UAS Landing Path and Obstacle Detection*,
Studies in Systems, Decision and Control 136,
https://doi.org/10.1007/978-3-319-73174-2

Papa, U., Russo, S., Lamboglia, A., Del Core, G., & Iannuzzo, G. (2017). Health Structure Monitoring for the Design of an Innovative UAS fixed wing through Inverse Finite Element Method (iFEM). *Aerospace Science and Technology*.

Papa, U., Russo, S. and Iannuzzo, G. (2017). Health Structure Monitoring for Aircraft and rotorcraft through Inverse Finite Element Method (iFEM). In *43rd European Rotorcraft Forum Proceedings*, Paper No. 575, Milan, Italy.

Papa, U., Fravolini, M. L., Del Core, G., Valigi, P., & Napolitano, M., R., (2017). Data-Driven Schemes for Robust Fault Detection of Air Data System Sensors. *IEEE Transactions on Control Systems Technology*.

Printed in the United States
By Bookmasters